METHUEN'S MONOGRAPHS
ON APPLIED PROBABILITY AND STATISTICS

*General Editor:* M. S. BARTLETT, F.R.S.

# POPULATION GENETICS

D0169091

# Population Genetics

W. J. EWENS

*Department of Mathematics,*
*La Trobe University, Melbourne*

METHUEN & CO LTD
11 New Fetter Lane · London E.C.4

*First published* 1969
© *Warren John Ewens* 1968
*Printed in Great Britain by*
*Willmer Brothers Limited*
*Birkenhead*

SBN 416 03160 9

Distributed in USA
by Barnes & Noble Inc.

# Contents

# CONTENTS

ix

# Preface

Population genetics is the mathematical investigation of the changes in the genetic structure of populations brought about by selection, mutation, inbreeding, migration, and other phenomena, together with those random changes deriving from chance events. These changes are the basic components of evolutionary progress, and an understanding of their effect is therefore necessary for an informed discussion of the reasons for and nature of evolution.

It would, however, be wrong to pretend that a mathematical theory, depending as it must on a large number of simplifying assumptions, should be accepted unreservedly and that its conclusions should be accepted uncritically. No-one would pretend that in the event of disagreement between observation and mathematical prediction, the discrepancy is due to anything other than the inadequacy of the mathematical treatment. The biological world is, of course, far too complex for the study of population genetics to be simply a branch of applied mathematics, so that while we are concerned here with the mathematical theory, I have tried to indicate which of our results should continue to apply in a context wider than that in which they are formally derived.

The difficulties involved in the joint discussions of mathematical and genetical problems are obvious enough. I have tried to aim this book rather more at the mathematician than at the geneticist, and for this reason a brief glossary of common genetical terms is included. On the other hand, it is not the aim of this book to stress mathematical techniques and results. Indeed, the mathematics is nowhere developed for its own sake, and proofs of several theorems have either been omitted or given in outline only, and the treatment of diffusion theory will be recognized as being heuristic and old-fashioned. It is hoped that the book may thus be of some use to the mathematically-inclined geneticist, and it is partly with this requirement in mind that it has been written.

In a short introductory book it is inevitable that some topics are excluded or treated very sketchily. The main omissions here are, firstly, the effects of inbreeding, and secondly, the theory of inheritance of continuous characters. An account of inbreeding would be more relevant to the theory of artificial (rather than natural) selection, and it is unlikely that the mathematical assumptions which are required to obtain a workable theory of inbreeding would apply even approximately in natural populations. An account of the theory of continuous characters has been omitted simply because of the space which an adequate discussion would require. A further restriction has been introduced by considering almost exclusively those populations whose size is effectively constant; this merely reflects the present state of knowledge. It is quite likely, as has been recently pointed out by Feller, that this restriction introduces a serious bias into many of our conclusions, and that it is important that results free of this assumption should be obtained.

I should like to take this opportunity of expressing my debt to Professors J. Gani, S. Karlin, E. J. Hannan and in particular P. A. P. Moran, for their help and advice over many years, and finally to thank Miss Helen Lambert for typing the manuscript.

Melbourne,                                                    W. J. Ewens
November, 1967

# CHAPTER 1

# The Hardy-Weinberg Law

## 1.1. The Hardy-Weinberg law

It does not often happen that the most important theorem in any subject is the easiest and most readily derived theorem for that subject, and the one which is first taught to students. Such, however, is the case in population genetics, and we shall now do no more than follow common practice by starting our considerations with such a theorem.

We consider a random-mating population* which is so large that we may ignore small chance variations in gene frequencies and treat all processes as being deterministic. Suppose that at any given locus only two alleles may occur, namely $A_1$ and $A_2$, and that individuals are diploid but monoecious, (i.e. can act as both male and female parents). Further, suppose that in any generation, the proportions of the three genotypes $A_1A_1$, $A_1A_2$ and $A_2A_2$ are $P$, $2Q$ and $R$ respectively.

Since random mating obtains, the frequency of matings of the type $A_1A_1 \times A_1A_1$ is $P^2$, that of $A_1A_1 \times A_1A_2$ is $4PQ$, and so on. We must now consider the outcomes of each of these matings. If the very small probability of mutation is ignored, elementary Mendelian rules indicate that the outcome of an $A_1A_1 \times A_1A_1$ mating must be $A_1A_1$, that half the $A_1A_1 \times A_1A_2$ matings will produce $A_1A_1$ offspring, the other half producing $A_1A_2$, with similar results for the remaining matings.

It follows that, since $A_1A_1$ offspring can be obtained only from $A_1A_1 \times A_1A_1$ matings (with frequency 1 for such matings), from $A_1A_1 \times A_1A_2$ matings (with frequency $\frac{1}{2}$ for such matings), and from $A_1A_2 \times A_1A_2$ matings (with frequency $\frac{1}{4}$ for such matings), and since the frequencies of these matings are $P^2$, $4PQ$, $4Q^2$, the frequency $P'$ of $A_1A_1$ in the following generation is

$$P' = P^2 + \tfrac{1}{2}(4PQ) + \tfrac{1}{4}(4Q^2) = (P+Q)^2. \tag{1.1}$$

* See glossary for definitions.

Similar considerations give

$$2Q' = \tfrac{1}{2}(4PQ) + \tfrac{1}{2}(4Q^2) + 2PR + \tfrac{1}{2}(4QR)$$
$$= 2(P+Q)(Q+R), \tag{1.2}$$
$$R' = \tfrac{1}{4}(4Q^2) + \tfrac{1}{2}(4QR) + R^2 = (Q+R)^2. \tag{1.3}$$

Note that in deriving these results we have assumed no selective differences between genotypes; that is we have assumed that the genotype of an individual affects neither his chance of surviving to produce offspring nor the number of such offspring.

The frequencies $P''$, $2Q''$ and $R''$ for the next generation are found by replacing $P'$, $2Q'$ and $R'$ by $P''$, $2Q''$ and $R''$ and $P$, $2Q$ and $R$ by $P'$, $2Q'$ and $R'$ in eqns. (1.1)–(1.3). Thus, for example,

$$P'' = (P'+Q')^2$$
$$= (P+Q)^2 \qquad \text{(Using (1.1) and (1.2))}$$
$$= P',$$

and similarly it is found that $Q'' = Q'$, $R'' = R'$. Thus the genotypic frequencies established by the second generation are maintained in the third generation and consequently in all subsequent generations. Note that frequencies having this property can be characterized as those satisfying the relation

$$(Q')^2 = P'R'. \tag{1.4}$$

Clearly, if this relation holds in the first generation, so that

$$Q^2 = PR, \tag{1.5}$$

then not only would there be no change in genotypic frequencies between the second and subsequent generations, but also these frequencies would be the same as those in the first generation. Populations for which eqn. (1.5) is true are said to have genotypic frequencies in Hardy-Weinberg form.

Since $P + 2Q + R = 1$, only two of the frequencies $P$, $2Q$ and $R$ are independent. If, further, eqn. (1.5) holds, only one frequency is independent. Examination of eqns. (1.1)–(1.3) shows that the most convenient quantity for independent consideration is $p = P+Q$, that is to say the frequency of the gene $A_1$. For convenience the notation $q = 1-p$ for the frequency of $A_2$ is often introduced, but this is not strictly necessary.

The above results may be summarized in the form of a theorem:

*Theorem* (Hardy-Weinberg). Under the assumptions stated, a population having genotypic frequencies $P$ (of $A_1A_1$), $2Q$ (of $A_1A_2$) and $R$ (of $A_2A_2$) achieves, after one generation of random mating, stable genotypic frequencies $p^2$, $2pq$, $q^2$ where $p = P+Q$ and $q = Q+R$. If the initial frequencies $P$, $2Q$, $R$ are already of the form $p^2$, $2pq$, $q^2$, then these frequencies are stable for all generations. This theorem was established independently by Hardy (1908) and Weinberg (1908); a particular case of the theorem was given by Pearson (1904).

It would be difficult to exaggerate the importance of this theorem. Unfortunately it is important for two different reasons, one purely technical, and concentration on the technical reason has sometimes tended to obscure its truly basic value. The technical point is that if, as we may reasonably assume, eqn. (1.5) is true, the mathematical behaviour of the population can be examined in terms of the single parameter $p$ rather than the vector $(P, Q)$; this is certainly a considerable convenience but is not fundamentally important. The really important part of the theorem lies in the stability behaviour; that is, if no external forces act, there is no intrinsic tendency for any variation present in the population (that is, any variation caused by the existence of the three different genotypes) to disappear.

The importance of this is based on the fact that evolution through natural selection can occur only if, within the population, there is variation upon which selective forces can act. Clearly variation, under a Mendelian system, tends to be maintained. (Of course, the action of selection itself often tends to destroy variation; this qualification is of some importance and we shall return to this point later.) Before the rediscovery of Mendelism in 1900, one commonly-held theory of inheritance was that the characteristics of offspring are a 'mixture' of the corresponding characteristics of parents, so that, for example, the skin colour of an offspring would be a mixture of the skin colours of the parents. It is easy to show that under random mating, the variation in the population for any characteristic would decrease under this theory by a factor of $\frac{1}{2}$ in each generation so that uniformity throughout the population would rapidly be obtained. This rapid rate is only slightly modified under reasonable non-

3

random mating. Thus no variation will usually exist upon which selection can act. Since, of course, we do not observe this uniformity of characteristics, further argument is required; but since variation from the expected 'mixture' can occur only by an additional factor which causes offspring not to resemble parents, and since to prevent rapid degeneration to uniformity this factor must have a very strong effect, it cannot reasonably be argued that selectively favoured parents produce offspring which closely resemble their parents and hence are themselves selectively favoured.

It is interesting to note that the consequences of the blending theory of inheritance were recognized by Darwin as a major obstacle to his theory of evolution through selection; it is the 'quantal' nature of Mendelian inheritance which completely removes this problem. This is all the more remarkable because it was often thought, in the decade following the rediscovery of Mendelian, that Darwinism and Mendelism were incompatible; results such as the Hardy-Weinberg theorem, and others considered later in this book, show that on the contrary the former practically depends on the latter for its operation.

## 1.2. Random union of gametes

The Hardy-Weinberg law was derived above under a number of simplifying assumptions, and in order to derive analogous laws under less restrictive assumptions, and to facilitate the mathematical arguments in general, we will now rederive the law in a more efficient way.

Any $A_1A_1$ parent will transmit an $A_1$ gene to his offspring. Any such gene is called, at this stage, a gamete; the union of two gametes forms a zygote or individual. Now the population considered in Section 1.1 produces $A_1$ gametes with frequency $P+Q$ and $A_2$ gametes with frequency $Q+R$; furthermore, random mating of individuals is equivalent to random union of gametes. Thus the frequency of $A_1A_1$ in the following generation is the frequency with which two gametes drawn independently are both $A_1$, namely $(P+Q)^2$. This establishes eqn. (1.1) and eqns. (1.2) and (1.3) follow similarly. The derivation of genotypic frequencies from the argument of random union of gametes will be used subsequently on a number of occasions.

## 1.3. Dioecious populations

We assumed in Section 1.1 that individuals are monoecious. While this assumption is of some independent interest, it was made mainly for convenience; to find how relevant the results derived from it are for other situations, some consideration must be given to populations where individuals are dioecious.

Suppose that the frequencies of $A_1A_1$, $A_1A_2$ and $A_2A_2$ among males are $P_M$, $2Q_M$ and $R_M$ and among females are $P_F$, $2Q_F$ and $R_F$. The gametic outputs from the two sexes are then $P_M + Q_M$ (of $A_1$) and $Q_M + R_M$ (of $A_2$) from males and $P_F + Q_F$ (of $A_1$) and $Q_F + R_F$ (of $A_2$) from females. The frequencies of the three genotypes (in both sexes) in the following generation are therefore $(P_M + Q_M)(P_F + Q_F)$, $(P_M + Q_M)(Q_F + R_F) + (P_F + Q_F)(Q_M + R_M)$, $(Q_M + R_M)(Q_F + R_F)$ respectively. The gametic output from this daughter generation is, for both sexes, $\frac{1}{2}(P_M + P_F + Q_M + Q_F)$ (of $A_1$) and $\frac{1}{2}(Q_M + Q_F + R_M + R_F)$ (of $A_2$). It is easy to show also that the genotypic frequencies in this following generation satisfy eqn. (1.5) for both sexes. Thus after one generation of random mating, the frequencies of the three genotypes are the same in both males and females, while a further generation of random mating ensures that these frequencies are in Hardy-Weinberg form.

Thus it is reasonable in many circumstances to ignore the dioecious nature of the population, and we shall indeed almost always do this, mentioning it only on occasions when special attention is necessary.

## 1.4. Sex-linked genes and multiple alleles

The theory of the preceding section does not apply when the genes in question are sex-linked, i.e. located on the sex chromosome. To analyse the behaviour for sex-linked genes, suppose that the male sex is heterogametic and that the initial frequencies are

| male | | female | | |
|---|---|---|---|---|
| $A_1$ | $A_2$ | $A_1A_1$ | $A_1A_2$ | $A_2A_2$ |
| $p_M$ | $q_M$ | $P_F$ | $2Q_F$ | $R_F$ . |

Consideration of the gametic output from each sex shows that in the following generation these frequencies become

| male | |
|---|---|
| $A_1$ | $A_2$ |
| $P_F + Q_F$ | $Q_F + R_F$ |

5

female

$$A_1A_1 \qquad\qquad A_1A_2 \qquad\qquad A_2A_2$$

$$p_M(P_F+Q_F) \quad q_M(P_F+Q_F)+p_M(Q_F+R_F) \quad q_M(Q_F+R_F).$$

The difference between the frequency of $A_1$ in males and the frequency of $A_1$ in females in the initial generation is

$$p_M-(P_F+Q_F), \tag{1.6}$$

while in the second generation this difference is

$$P_F+Q_F-\{p_M(P_F+Q_F)+\tfrac{1}{2}q_M(P_F+Q_F)+\tfrac{1}{2}p_M(Q_F+R_F)\}$$
$$= -\tfrac{1}{2}\{p_M-(P_F+Q_F)\}, \tag{1.7}$$

which in absolute value is half of (1.6). Clearly, with succeeding generations, this difference rapidly approaches zero. If, then, it is assumed that initially $p_M = P_F+Q_F = p$ (say), the above theory shows that in one generation the frequencies

| male | | female | | |
|:---:|:---:|:---:|:---:|:---:|
| $A_1$ | $A_2$ | $A_1A_1$ | $A_1A_2$ | $A_2A_2$ |
| $p$ | $q$ | $p^2$ | $2pq$ | $q^2$ |

are attained, and that these frequencies are unaltered in subsequent generations. For arbitrary initial values of $P_F$ and $Q_F$, this does not happen, although a very rapid convergence to such an equilibrium state occurs. In any event, the important part of the Hardy-Weinberg law relating to essential stability of genotypic frequencies still stands.

The Hardy-Weinberg law can be extended immediately to the case where more than two types of genes are allowed at the locus in question. If alleles $A_1, \ldots, A_k$ occurs with frequencies $p_1, \ldots, p_k$, then after one generation of random mating the frequency of $A_iA_i$ is $p_i^2$, while that of $A_iA_j(i \neq j)$ is $2p_ip_j$; in subsequent generations these frequencies are unaltered. The proofs of these statements follow immediately by considering gamete frequencies, and are omitted; again it is clear that genotypic frequencies are essentially stable.

## 1.5. Miscellaneous results

In this section some elementary considerations derived from the Hardy-Weinberg law will be examined.

In a number of cases, the gene $A_1$ is *dominant* to $A_2$; that is, $A_1A_1$

individuals are indistinguishable from $A_1A_2$. A common fallacy in such a situation is to suppose that such dominance 'spreads' and that eventually all individuals will be indistinguishable. Such is not the case, for the stable frequencies derived in Section 1.1 apply irrespective of the existence of dominance; what is gained in the frequency of dominant individuals by mating of $A_1A_1$ with $A_2A_2$ and of $A_1A_2$ with $A_2A_2$ is exactly counterbalanced by the loss in frequency through matings of $A_1A_2$ with $A_1A_2$ and of $A_1A_2$ with $A_2A_2$.

A second consequence of the Hardy-Weinberg law is that if $A_2$ is recessive to $A_1$ and has small frequency, we shall rarely observe recessive individuals. Further, the parents of recessives will usually both be heterozygotes. For example, if the frequency $q$ of $A_2$ is 0·001, then the frequency of $A_2A_2$ is 0·000001. The frequency with which an $A_2A_2$ individual has both parents $A_1A_2$ may be found from the fact that the parents of an $A_2$ individual must both be $-A_2$, where the unknown gene is either $A_1$ or $A_2$. The frequency with which both unknown genes are $A_1$ is $(0·999)^2 = 0·998001$. This indicates that the attempted removal of a rare recessive gene by removal of recessives $A_2A_2$ will have but a minor effect; later on the rate at which such removal will decrease the frequency of $A_2$ will be considered.

Finally, we remark that the Hardy-Weinberg law has been derived here under the assumption that generations do not overlap. Thus if this assumption does not hold, the law itself may not hold. For example, suppose as a continuous time analogue to the above that in a small time $dt$ a fraction $dt$ of the population dies and is replaced, by random sampling, from the population at large. Under this system the frequency $p$ of $A_1$ does not change with time, but if $P(t)$ is the frequency $A_1A_1$ at time $t$, then

$$P(t+dt) = P(t)\,(1-dt)+p^2\,dt.$$

Passing to the limit in this equation,

$$\frac{dP(t)}{dt} = -P(t)+p^2,$$

so that

$$P(t) = \{P(0)-p^2\}\exp(-t)+p^2.$$

Clearly a population initially in Hardy-Weinberg equilibrium will remain in equilibrium, but for non-equilibrium populations, the equilibrium state is approached asymptotically (and rapidly). It is clear that the important conclusions derived from the Hardy-Weinberg law remain unchanged. For a more complete discussion of this and similar problems, see Moran (1962, p. 23).

### 1.6. The effect of selection

The results given above have been derived under the assumption that no selective differences exist between the three genotypes $A_1A_1$, $A_1A_2$ and $A_2A_2$. In attempting to discuss the effect of selection one immediately comes up against the problem that selective values are not properties of genes; they are rather properties of individuals (i.e. of the whole interacting collection of genes which an individual has), and then refer properly only to a given environment. Thus it may, and often does, happen that a gene which is selectively advantageous against one genetic background is disadvantageous against another. It will be shown later that such interaction effects can have major evolutionary consequences, and that it appears difficult even to define a concept of 'independence' of loci. For the moment, we make the rough approximation that selective differences depend on the genotype at a given single locus; despite the above remarks, this approximation leads to a number of valuable results.

To be definite, suppose that if, at the time of conception of any generation, the frequencies of the genotypes are $P$, $2Q$, $R$, then these genotypes contribute gametes to form the individuals of the following generation in the proportions $w_{11}P:2w_{12}Q:w_{22}R$. (Note that the population is being considered at the time of formation of zygotes from the gametes of the previous generation. When selective differences exist, this is the only time when Hardy-Weinberg proportions strictly apply; later on, when considering finite populations, the population will be counted at the age of sexual maturity.) The differential reproduction rates may be due to several causes, including in particular different survival rates and different offspring distributions. The quantities $w_{11}$, $w_{12}$, and $w_{22}$ will be called the 'fitnesses' of the three genotypes, and when these are not all equal, selective forces are operating and genotypic frequencies will usually change from one generation to the next.

With the fitnesses given above, the frequencies of the various genotypes in the following generation now satisfy the equation

8

$P':2Q':R'$

$$= (w_{11}P + w_{12}Q)^2 : 2(w_{11}P + w_{12}Q)(w_{12}Q + w_{22}R) : (w_{12}Q + w_{22}R)^2$$

$$= (p')^2 : 2p'q' : (q')^2, \tag{1.8}$$

where

$$p' = (w_{11}P + w_{12}Q)/(w_{11}P + 2w_{12}Q + w_{22}R). \tag{1.9}$$

Clearly, after one generation of random mating, Hardy-Weinberg proportions are achieved. In the next generation, the same arguments show that

$$P'':2Q'':R'' = (p'')^2 : 2p''q'' : (q'')^2$$

$$= \{w_{11}(p')^2 + w_{12}p'q'\}^2 : 2\{w_{11}(p')^2 + w_{12}p'q'\}$$

$$\times \{w_{12}p'q' + w_{22}(q')^2\} : \{w_{12}p'q' + w_{22}(q')^2\}^2. \tag{1.10}$$

It follows that the equation $p'' = p'$ no longer holds in general. Thus while genotype frequencies settle down immediately to Hardy-Weinberg form, the more important part of the Hardy-Weinberg theorem relating to constancy of genotypic frequencies no longer holds. We shall examine some consequences of this conclusion in the next chapter.

# CHAPTER 2

# Selection and Mutation

## 2.1. Changes in gene frequency

One of the conclusions of the previous chapter was that the existence of selective differences among genotypes generally leads to changes in gene frequencies. Assuming Hardy-Weinberg proportions, the frequency $p'$ of $A_1$ in any generation is related to its frequency $p$ in the previous generation by

$$p' = \frac{w_{11}p^2 + w_{12}pq}{w_{11}p^2 + 2w_{12}pq + w_{22}q^2}, \tag{2.1}$$

where the $w_{ij}$ are defined in Section 1.6. The change $\Delta p$ in the frequency of $A_1$ is thus

$$\Delta p = pq\, \frac{w_{11}p + w_{12}(1-2p) - w_{22}q}{w_{11}p^2 + 2w_{12}pq + w_{22}q^2}. \tag{2.2}$$

Clearly $\Delta p = 0$ whenever $p = 0$ or $p = 1$, corresponding to fixation of $A_2$ or $A_1$. $\Delta p$ is also zero when

$$p = p^* = \frac{w_{12} - w_{22}}{(w_{12} - w_{22}) + (w_{12} - w_{11})}. \tag{2.3}$$

Now p* will lie in $(0, 1)$ only when $w_{12}$ is less than both $w_{11}$ and $w_{22}$, or when $w_{12}$ exceeds both $w_{11}$ and $w_{22}$. When the former condition holds, the equilibrium at the point $p = p^*$ is unstable; since in finite populations small deviations are bound to occur, the equilibrium point $p^*$ is of little interest in this case. On the other hand, when $w_{12}$ exceeds both $w_{11}$ and $w_{22}$, the equilibrium at $p^*$ is stable, so that small deviations of $p$ from $p^*$ tend to be restored in successive generations. Indeed $\Delta p$ is then positive for all $p$ in $(0, p^*)$ and negative for all $p$ in $(p^*, 1)$. We conclude that the necessary and sufficient condition that, under selective pressures only, there exist a stable equilibrium of gene frequency in $(0, 1)$ is that the heterozygote have larger fitness than both homozygotes. This most important fact was established by Fisher (1922).

Perhaps the best-known example of this situation in man concerns the phenomenon of sickle-cell anaemia. The maintenance of high frequencies for both the sickle cell gene and its normal allele in certain East African tribes appears to be due to a selective advantage of heterozygotes brought about by the increased resistance of such heterozygotes to malaria. For a mathematical investigation of this problem, see Smith (1955).

It is of some interest to note that eqn. (2.3) shows that at most one stable internal equilibrium point can exist under the assumed conditions. This is no longer true when more general conditions are allowed. For example, in dioecious populations, when the two sexes have different sets of fitnesses for the three genotypes, it is possible that two stable internal equilibrium points exist. The conditions that this is the case, together with a numerical example, have been given by Owen (1953), (see also Li (1963)). For a more complicated model, see Bodmer (1964).

If an arbitrary number $m$ of alleles is allowed at the locus in question, the problem of the existence of internal equilibrium points is more complex. Thus if the fitness of $A_iA_j$ is $w_{ij}(i,j = 1,2, \ldots m)$, there can be at most one equilibrium point for which each $p_i > 0$. This equilibrium point, if it exists, will be stable if the matrix $W = \{w_{ij}\}$ has exactly one positive eigenvalue and at least one negative eigenvalue. In this case the system moves, irrespective of the initial gene frequencies, towards this equilibrium point. There may be additional equilibrium points in which various sets of genes are absent, but for any given set of genes present there can be at most one equilibrium point. Further, if $W$ has $k$ positive eigenvalues, at least $k-1$ genes must die out (i.e. have frequency asymptotically approaching zero) before equilibrium is attained.

If there is no stable internal (i.e. each $p_i > 0$) equilibrium point, the system will move to one or other of the boundary (i.e. some of the $p_i = 0$) equilibrium points. The conditions on the $w_{ij}$ and the $p_i$ which determine which of these is approached are unknown. These and further results are proved by Kingman (1961a), (see also Mandel (1959)); the details of the proofs will be omitted here.

Returning to the case of two alleles, if $w_{12}$ lies between $w_{11}$ and $w_{22}$, the frequency $p$ will increase or decrease steadily, depending on

11

whether $w_{11} > w_{22}$ or $w_{11} < w_{22}$. For example, if $w_{12} = \frac{1}{2}(w_{11}+w_{22})$, then

$$\Delta p = \tfrac{1}{2}pq \, \frac{w_{11}-w_{22}}{w_{11}p+w_{22}q}$$

If $w_{11} > w_{22}$, this is positive for all $p$ in $(0,1)$, so that the frequency of $A_1$ steadily increases; when $w_{11} < w_{22}$, the frequency of $A_1$ steadily decreases.

Since $p$ will generally change from one generation to the next, it is of some interest to find the time required for any given change in gene frequency. The times so required are relevant in particular to the discussion of the rate at which genetic variation is destroyed by the action of selection. If time is measured in units of one generation, eqn. (2.2) can be replaced, to a sufficiently close approximation, by the equation

$$\frac{dp}{dt} = pq \, \frac{w_{11}p+w_{12}(1-2p)-w_{22}q}{w_{11}p^2+2w_{12}pq+w_{22}q^2}. \tag{2.4}$$

It follows that the time $t(p_1, p_2)$ required for $p$ to move from $p_1$ to $p_2$ is

$$t(p_1,p_2) = \int_{p_1}^{p_2} \frac{w_{11}p^2+2w_{12}pq+w_{22}q^2}{pq\{w_{11}p+w_{12}(1-2p)-w_{22}q\}} \, dp. \tag{2.5}$$

(Naturally this equation applies only for cases where, starting from $p_1$, the frequency of $A_1$ will eventually reach $p_2$.)

It will often occur that the fitnesses $w_{11}$, $w_{12}$ and $w_{22}$ are all close to unity, so that since in any event only ratios of fitnesses are relevant, it is convenient to write

$$w_{11} = 1+s_1, \; w_{12} = 1+s_2, \; w_{22} = 1, \tag{2.6}$$

where $s_1$ and $s_2$ are small. In this case, eqn. (2.5) can be replaced, to a sufficiently close approximation, by

$$t(p_1,p_2) = \int_{p_1}^{p_2} \frac{dp}{pq\{s_1p+s_2(1-2p)\}}. \tag{2.7}$$

Once $p$ approaches unity, the time required for even small changes in $p$ will be large, due to the small term $q$ in the denominator of the integrand. This behaviour is even more marked when $s_1 = s_2$ (i.e. fitness of $A_1A_1$ = fitness of $A_1A_2$), for then eqn. (2.7) reduces to

$$t(p_1, p_2) = \int_{p_1}^{p_2} \frac{dp}{s_1 p q^2}. \tag{2.8}$$

Here the rate of increase of $p$, once $p$ is near unity, is extremely slow, due to the extreme rarity of homozygotes $A_2A_2$ (against whom selection is operating). In the important particular case $s_2 = \frac{1}{2}s_1$, eqn. (2.7) assumes the form

$$t(p_1, p_2) = \int_{p_1}^{p_2} \frac{dp}{s_2 p q}. \tag{2.9}$$

From these equations it is possible to evaluate the times required for various changes in frequency for various values of $s_1$ and $s_2$, and some representative values are shown below in Table 2.1. Note that the times shown are rather long; thus while, in the cases considered, selection acts ultimately to destroy variation, the time required to do this is usually quite long, and certainly is far longer than the time required under any blending theory of inheritance.

TABLE 2.1. Times required (in terms of generations) for indicated changes in gene frequency for various fitness values.

| $s_1 = 0.001$, $s_2 = 0.0005$ | | | | | |
|---|---|---|---|---|---|
| Range | 0.1–0.99 | 0.99–0.999 | 0.999–0.9999 | 0.9999–0.99999 | 0.99999–0.999999 |
| Time | 13,600 | 4,600 | 4,600 | 4,600 | 4,600 |

| $s_1 = s_2 = 0.001$ | | | | | |
|---|---|---|---|---|---|
| Range | 0.1–0.8 | 0.8–0.95 | 0.95–0.98 | 0.98–0.99 | 0.99–0.999 | 0.999–0.99999 |
| Time | 7,500 | 21,800 | 25,700 | 50,700 | 902,300 | 99,000,000 |

We shall return to this table on several occasions later. For analogous tables which consider more complicated situations allowing for inbreeding, different sets of fitnesses for the two sexes, and so on, see Haldane (1924, 1926, 1927a, 1927b, 1930a, 1930b, 1932a, 1932b).

## 2.2. The mean fitness of the population

Since at any time the frequencies of the various genotypes are of the form $p^2$, $2pq$, $q^2$, while the corresponding fitnesses are $w_{11}$, $w_{12}$, $w_{22}$, the mean fitness $W$ of the population at the time in question is

$$W = w_{11}p^2 + 2w_{12}pq + w_{22}q^2. \tag{2.10}$$

If $w_{11} > w_{12} > w_{22}$, this is a montonically increasing function of $p$; if $w_{11} < w_{12} < w_{22}$ it decreases monotonically with $p$. Otherwise, $W$ has a turning point when $dW/dp = 0$; this occurs where $p = p^*$ (defined by eqn. (2.3)). If $w_{12} > w_{11}$, $w_{12} > w_{22}$, $W$ achieves a maximum at $p^*$, while if $w_{12} < w_{11}$, $w_{12} < w_{22}$, $W$ achieves a minimum at $p^*$. It is clear from the discussion of the previous section that $p$ always moves towards local maxima of $W$, and that, because of the form of eqn. (2.10), this means that the mean fitness of any generation always exceeds (or at worst is equal to) that of the previous generation. The generalizations of this result will be considered in Chapter 3; for the moment we note merely that it accords with intuitive expectations.

## 2.3. The effect of mutation

So far we have ignored completely the possibility that any gene can mutate. Since natural mutation rates are usually quite small (of the order $10^{-5}$ or $10^{-6}$) this may, in some circumstances, be permissible; however mutation is a key factor in evolution and it is necessary to examine quantitatively some of its consequences. For the moment only a preliminary investigation, concerned specifically with the theory considered above, will be given; a more complete examination must take stochastic fluctuations into account and will be considered later.

Suppose that the mutation rate $A_1 \rightarrow A_2$ is $u$, with counter-mutation $A_2 \rightarrow A_1$ at rate $v$. Such mutation will mean that neither $A_1$ nor $A_2$ can be permanently fixed in the population, and that there must be stable internal equilibrium points.

Considering mutation pressures only, the change in the frequency of $A_1$ between successive generations is

$$\Delta p = -up + v(1-p). \tag{2.11}$$

This is zero when $p = v/(u+v)$ which, under mutation pressure only, is a point of stable equilibrium.

Before considering, as we must, the joint effect of selection and mutation, it is worth while demonstrating the extremely slow rate at which gene frequencies change under mutation pressure alone. If $u = 0$, the frequency of $A_1$ will steadily increase, due to persistent mutation from $A_2$ to $A_1$. To a sufficient approximation, eqn. (2.11) shows that the rate of increase of the frequency of $A_1$ is

$$\frac{dp}{dt} = v(1-p).$$

Thus if $t(p_1, p_2)$ is the time required for $p$ to increase from $p_1$ to $p_2$,

$$t(p_1, p_2) = \int_{p_1}^{p_2} \frac{dp}{v(1-p)}. \qquad (2.12)$$

Taking, for example, $v = 5 \times 10^{-6}$, eqn. (2.12) shows that the time required for $p$ to increase from 0·1 to 0·9 is $1 \cdot 1 \times 10^6$ generations. The same time is required for an increase in frequency from 0·9 to 0·99. Clearly under normal mutation rates, even if reverse mutation is not allowed, only extremely slow changes in gene frequencies will occur.

We turn now to the joint consideration of mutation and selection. To an extremely close approximation the changes in gene frequency due to these two factors acting together may be taken as the sum of the changes in gene frequency which occur when these factors act separately; this implies that

$$\Delta p = pq \frac{w_{11}p + w_{12}(1-2p) - w_{22}q}{w_{11}p^2 + 2w_{12}pq + w_{22}q^2} + v(1-p) - up. \qquad (2.13)$$

To make further progress it is necessary to make some assumptions about the relative values of the $w_{ij}$. As mentioned above, these will often be rather close to each other, differing by a factor of order 1 per cent. In this case, eqn. (2.13) may be replaced, in the notation of Section 2.1, by

$$\Delta p = pq\{s_1 p + s_2(1-2p)\} + v(1-p) - up. \qquad (2.14)$$

The form of this equation shows that there is always at least one solution of the equation

$$\Delta p = 0 \qquad (2.15)$$

in the range $(0,1)$. There may, in some cases, be two or even three

15

solutions of eqn. (2.15) in this range, but in the latter case at most two are stable.

We shall often be concerned with the case where $s_1$ and $s_2$, although small, are nevertheless considerably larger than $u$ and $v$. We shall distinguish three special cases of the sort of behaviour which can occur for various values of the selection parameters. Firstly, when $s_2 > s_1 > 0$, there exists in the absence of mutation a stable internal equilibrium point, which is only slightly shifted when mutation is allowed. Thus if $s_2 = 0\cdot03$, $s_1 = 0\cdot01$, the equilibrium point is $p = 0\cdot6$. If now $u = v = 10^{-5}$, the equilibrium point is moved to $0\cdot5983$, so that the mutation has a negligible effect. The second case occurs when $|s_1| > |s_2|$; here a stable equilibrium, maintained by a balance between mutation and selection, will exist. This equilibrium is near $p = 1$ if $s_2 > 0$ and is near $p = 0$ if $s_2 < 0$, and its nature can be exemplified by considering the case $s_2 > 0$. To take a numerical example, if $s_1 = 0\cdot02$, $s_2 = 0\cdot01$, $u = v = 10^{-5}$, then the solution of eqn. (2.15) is approximately $p = 0\cdot999$. In general, the equilibrium frequency of $A_1$ in such cases is found by replacing eqn. (2.14) by

$$\Delta p = q\{s_1 - s_2\} - u,$$

in which case the solution of eqn. (2.15) is

$$p = 1 - \{u/(s_1 - s_2)\}. \tag{2.16}$$

When $s_1 < s_2 < 0$, similar reasoning shows that at equilibrium,

$$p \approx v/|s_2|. \tag{2.17}$$

Finally, when $s_1 = s_2 > 0$, the equilibrium frequency is

$$p \approx 1 - \sqrt{(u/s_2)}. \tag{2.18}$$

This is generally somewhat lower than (2.16); thus if $s_1 = s_2 = 0\cdot01$, $u = 10^{-5}$, the equilibrium frequency of $A_1$ is $0\cdot9684$. The reason for such behaviour is obvious. $A_2$ suffers selective disadvantage only when in homozygotes $A_2 A_2$; for $p$ near unity the frequency of such homozygotes is so small that the great majority of $A_2$ genes occur in heterozygotes, which suffer no selective disadvantage. This leads to a considerable increase in the frequency of $A_2$. It might be remarked in conclusion that this phenomenon is observed frequently in nature, since many rare unfavoured genes appear to be recessive.

## 2.4. Genetic loads

When there is a stable equilibrium gene frequency in the interval $(0,1)$, due either to selective advantage of the heterozygote or to mutation, the mean fitness of the population will be less than the fitness of the most fit genotype. In such cases we shall say that the population is carrying a 'genetic load'; in the two cases mentioned above, this is called respectively a segregational load and a mutational load. It is perhaps unfortunate that the word 'load' is used in this context, since it is in many ways desirable that such a load should exist. Thus the presence of a mutational load, for example, ensures that should changes in environment occur which make the mutant more fit than the wild type, the population has the capacity to respond to such a change through an increase in the frequency of the mutant. The term 'load' is, however, now used so widely that we shall employ it and concentrate attention on a quantitative treatment of the concept.

If the maximum of the $w_{ij}$ is denoted $w_{max}$, several reasonable mathematical definitions for the genetic load can be made. Crow (1958) has used the expression $(w_{max} - W)/w_{max}$; in this book we shall use the closely associated expression

$$L = (w_{max} - W)/W. \qquad (2.19)$$

There are no very strong reasons for using eqn. (2.19) in preference to Crow's expression; we use eqn. (2.19) mainly because it indicates immediately how much in excess of unity the fitness of the most fit genotype must be to preserve a stable population size (i.e. $W = 1$).

Suppose that, using the notation (2.6), $|s_1| \geqq |s_2| > 0$. Then heterozygosity is maintained by mutation, and use of eqns. (2.16) and (2.17) shows that when $|s_1| > |s_2|$, the genetic load is approximately twice the mutation rate of the more favoured gene. When $s_1 = s_2$, the genetic load is about half this amount. In either case the genetic load is very small. Fraser (1962) has discussed the relevance of results such as these to Man.

A segregational load will exist if $s_2 > s_1 > 0$. In this case, $p = s_2/(2s_2 - s_1)$ and elementary operations show that if the $s_i$ are small, the segregational load is approximately $s_2(s_2 - s_1)/(2s_2 - s_1)$. This will usually be rather larger than the mutational load. Indeed, one of the main problems of population genetics today is to enquire whether the large amounts of heterozygosity observed in many populations

17

can be accounted for by selective advantage of heterozygotes, for if so, and if the number of heterozygous loci involved is large, the population would be carrying a very heavy genetic load. This problem is discussed at length by Lewontin and Hubby (1966).

In subsequent chapters, several ways in which the genetic load can affect the behaviour of populations will be discussed.

# CHAPTER 3

# The Fundamental Theorem of Natural Selection

The essence of the theory of evolution through selection is that in any population there will exist genetic variation between individuals and that those genotypes which are better suited to the environment than others will contribute rather more than their fair share of off-spring to the following generation. Thus the genetical make-up of the following generation will differ somewhat from that of the parent generation, leading to substantial changes over large numbers of generations.

Such evolution depends on the existence of genetical variation in the population, so that it might be expected that the greater the variation, the greater will be the changes which occur. Further, it appears that in some sense the process leads to an 'improvement' in the population. The theory which has so far been developed allows a more precise quantitative examination of these intuitive notions.

## 3.1. Random-mating populations

Consider firstly the case where only two alleles $A_1$ and $A_2$ are allowed at the locus in question. Using the notation developed in the previous two chapters, if the frequency of $A_1$ in any generation is $p$, and that of $A_2$ is $q = 1-p$, the frequency $p'$ of $A_1$ in the following generation is

$$p' = (w_{11}p^2 + w_{12}pq)/W, \tag{3.1}$$

where $W$ is given by eqn. (2.10).

The mean fitness $W'$ of the population in the second generation is

$$W' = w_{11}(p')^2 + 2w_{12}p'q' + w_{22}(q')^2, \tag{3.2}$$

and the increase $\Delta W$ in mean fitness between the two generations is

$$\Delta W = w_{11}\{(p')^2 - p^2\} + 2w_{12}\{p'q' - pq\} + w_{22}\{(q')^2 - q^2\}$$
$$= (p' - p)\{w_{11}(p' + p) + 2w_{12}(1 - p - p') + w_{22}(p + p' - 2)\}. \quad (3.3)$$

Writing $\Delta p = p' - p$ and using the expression (3.1) for $p'$, it is possible after some manipulation to reduce eqn. (3.3) to the form

$$\Delta W = (\Delta p)^2\{w_{11} - 2w_{12} + w_{22} + 2W(pq)^{-1}\}$$
$$= 2pq\{w_{11}p + w_{12}(1 - 2p) - w_{22}q\}^2$$
$$\{w_{11}p^2 + (w_{12} + \tfrac{1}{2}w_{11} + \tfrac{1}{2}w_{22})pq + w_{22}q^2\}W^{-2}. \quad (3.4)$$

Clearly $\Delta W$ is always non-negative, so that we can conclude that gene frequencies move, under natural selection, in such a way as to increase, or at worst maintain, the mean fitness of the population.

If, furthermore, the $w_{ij}$ are close to unity, eqn. (3.4) may be approximated by

$$\Delta W = 2pq\{w_{11}p + w_{12}(1 - 2p) - w_{22}q\}^2 = 2pq\{E_1 - E_2\}^2, \quad (3.5)$$

where
$$E_1 = w_{11}p + w_{12}q - W \quad (3.6)$$
and
$$E_2 = w_{12}p + w_{22}q - W. \quad (3.7)$$

The quantities $E_1$ and $E_2$ can be given the following interpretation. If the heterozygotes $A_1A_2$ are divided into two halves, the first half going to form a group with the homozygotes $A_1A_1$ and the second half going to form a group with the homozygotes $A_2A_2$, then $E_1$ and $E_2$ represent respectively the deviations from the mean fitness of the population of the mean fitnesses of the two groups. The quantity $E_1 - E_2$ is called the 'average excess' of $A_1$ (Fisher, 1930).

It was remarked earlier than it is expected that $\Delta W$ will be related in some way to the variation in fitness in the population. This possibility is now investigated more closely. It is clear from eqn. (3.5) that if $E_1 = E_2$, that is to say if

$$p = p^* = \frac{w_{12} - w_{22}}{2w_{12} - w_{11} - w^{22}}, \quad (3.8)$$

then $\Delta W = 0$. In the case when $w_{12}$ exceeds both $w_{11}$ and $w_{22}$, $p^*$ is, as we know, a stable equilibrium point, so that it could have been anticipated that $W$ will remain unchanged when $p = p^*$. But the total variance in fitness, namely

$$\sigma^2 = w_{11}^2 p^2 + 2w_{12}^2 pq + w_{22}^2 q^2 - W^2 \qquad (3.9)$$

will be positive when $w_{12}$ exceeds both $w_{11}$ and $w_{22}$. Thus if $\Delta W$ is to have some interpretation as a variance, it can only be as some component of the total variance of fitness.

To guide us in trying to isolate some component of $\sigma^2$ which is related to $\Delta W$, it is useful to consider the particular case $p = q = \frac{1}{2}$, $w_{11} = w_{22} = 1$, $w_{12} = 1+c$. Different values of $c$ lead to different values of $\sigma^2$, but irrespective of $c$ it is always true that $\Delta W = 0$. This suggests that it would be useful to isolate some component of $\sigma^2$ which is zero irrespective of the value of $c$. A component of $\sigma^2$ fulfilling this requirement is that part of the total variance which is removed by fitting a weighted least-squares regression line to the fitnesses $w_{ij}$. It is therefore reasonable to consider generally the effect of fitting such a line. Any regression line will yield fitness value $W+2x$, $W+x+y$, $W+2y$ for $A_1A_1$, $A_1A_2$, and $A_2A_2$, and the least-squares regression line is that line for which

$$D = p^2(w_{11} - W - 2x)^2 + 2pq(w_{12} - W - x - y)^2 + q^2(w_{22} - W - 2y)^2$$

is minimized with respect to variation in $x$ and $y$. The required solutions for $x$ and $y$ are

$$x = E_1 - W, \quad y = E_2 - W, \qquad (3.10)$$

where $E_1$ and $E_2$ are defined by eqns. (3.6) and (3.7). Standard regression theory shows that the sum of squares removed by fitting this least-squares line is

$$\sigma_A^2 = 2pq(E_1 - E_2)^2, \qquad (3.11)$$

which is identical to eqn. (3.5). Thus to the order of approximation used the increase in mean fitness of the population is identical to that part of the total genetic variance which can be accounted for by fitting a weighted least-squares regression line to the fitnesses. For this reason, this component is called the 'additive' part of the total variance.

We now consider the extension of these results to an arbitrary number $m$ of alleles. If the frequency of $A_i$ is $p_i$ and the fitness of $A_iA_j$ is $w_{ij}$, the mean fitness $W$ of the population is

$$W = \sum_i \sum_j w_{ij} \, p_i \, p_j. \qquad (3.12)$$

The frequency $p_i'$ of $A_i$ in the following generation is then

$$p_i' = W^{-1} \sum_j w_{ij} \, p_i \, p_j. \qquad (3.13)$$

It follows that the mean fitness $W'$ of the population in the following generation is given by

$$\begin{aligned}
W' &= \sum \sum w_{ij} \, p_i' \, p_j' \\
&= W^{-2} \{ \sum_i \sum_j \sum_k \sum_l w_{ij} \, w_{ik} \, w_{lj} \, p_i \, p_j \, p_k \, p_l \},
\end{aligned} \qquad (3.14)$$

so that

$$\Delta W = W^{-2} \{ \sum_i \sum_j \sum_k \sum_l w_{ij} \, w_{ik} \, w_{lj} \, p_i \, p_j \, p_k \, p_l - W^3 \}. \qquad (3.15)$$

We expect from our previous result that this quantity will be positive, and prove that this is the case in the following way. Firstly, if $p_i \geq 0$, $\sum p_i = 1$ and $b_i \geq 0$, then for $n \geq 1$ the inequality

$$\sum p_i \, b_i^n \geq (\sum p_i \, b_i)^n \qquad (3.16)$$

will always hold. This assertion is easily proved by using the convexity property of the function $b^n$ for $n \geq 1$. Then

$$\sum \sum \sum w_{ij} \, w_{ik} \, w_{lj} \, p_i \, p_j \, p_k \, p_l$$

$$= \tfrac{1}{2} \sum_i \sum_j \sum_k w_{ij} \, w_{ik} \, (\sum_l w_{lj} \, p_l + \sum_m w_{mk} \, p_m) \, p_i \, p_j \, p_k$$

$$\geq \sum_i \sum_j \sum_k w_{ij} \, w_{ik} \, (\sum_l \sum_m w_{lj} \, w_{mk} \, p_l \, p_m)^{\frac{1}{2}} \, p_i \, p_j \, p_k \qquad (3.17)$$

$$= \sum_i p_i \{ \sum_j w_{ij} \, p_j \, (\sum_l w_{lj} \, p_l)^{\frac{1}{2}} \}^2$$

$$\geq [\sum_i p_i \sum_j w_{ij} \, p_j \, (\sum_l w_{lj} \, p_l)^{\frac{1}{2}}]^2 \qquad (3.18)$$

$$= [\sum_i p_j \, (\sum_l w_{lj} \, p_l)^{3/2}]^2$$

$$\geq [\sum_i \sum_j w_{ij} \, p_i \, p_j]^3 = W^3. \qquad (3.19)$$

In this sequence of steps the inequalities (3.18) and (3.19) are justified from (3.16), while (3.17) is justified by the inequality $\frac{1}{2}(x+y) \geqq \sqrt{(xy)}$ for positive quantities $x, y$. This sequence of inequalities proves $W' \geqq W$, but also makes clear (Kingman, 1961b) the condition under which $W' = W$. Each inequality will be an equality if and only if

$$\sum_j w_{ij} p_j = c, \qquad (i = 1, 2, \ldots m) \qquad (3.20)$$

where $c$ is a constant independent of $i$. Multiplying the $i^{th}$ equation in (3.20) by $p_i$ and adding, it appears that $c$ must be identical to $W$. Further, equating $p_i'$ to $p_i$ in eqn. (3.13) yields

$$\sum_j w_{ij} p_j = W. \qquad (3.21)$$

We conclude that fitness increases monotonically, being constant for internal (i.e. each $p_i > 0$) values of $p_i$ only for that set of $p_i$ satisfying eqn. (3.21). This set of values yields a point of stable equilibrium. On the other hand, the $p_i$'s neet not approach this particular equilibrium point; they can approach a 'boundary' equilibrium point where one or more of the $p_i$'s are zero. The problem of determining 'domains of attraction', that is, of determining the set of initial $p_i$'s which lead eventually to a given equilibrium point, remains unsolved.

While we are primarily concerned with changes of mean fitness, it is of some interest to note that changes in gene frequency also obey an optimum property; this has been demonstrated by Kimura (1958). Suppose that the genotypic frequencies are in Hardy-Weinberg form, but not necessarily at a point of stable equilibrium. Suppose further that at this point, $W = 1$. Then from eqn. (3.13),

$$\Delta p_i = p_i \{ \sum_j w_{ij} p_j - 1 \},$$

so that

$$\sum_{i=1}^m \frac{(\Delta p_i)^2}{p_i} = \sum_{i=1}^m p_i \{ \sum_j w_{ij} p_j - 1 \}^2. \qquad (3.22)$$

It is easy to show that the right-hand side in eqn. (3.22) is $\frac{1}{2}\sigma_A^2$, where $\sigma_A^2$ is the additive portion of the genetic variance (defined by an extension of eqn. (3.11)).

Suppose now that in the following generation the frequency of $A_i$ is $p_i + \delta_i$. Thus the change in mean fitness of the population is

$$\Delta W = \sum \sum w_{ij} \{(p_i + \delta_i)(p_j + \delta_j) - p_i p_j\}$$

$$= 2 \sum_i \{\sum_j w_{ij} p_j\}\delta_i, \quad \text{(ignoring terms of order } \delta_i{}^2\text{)}$$

$$= 2 \sum_i w_i \cdot \delta_i, \text{ say.}$$

Then Kimura has shown that of all deviations $\{\delta_i\}$ satisfying

$$\sum \delta_i = 0, \quad \sum \frac{\delta_i{}^2}{p_i} = \tfrac{1}{2}\sigma_A{}^2,$$

the set maximizing $\sum w_i \cdot \delta_i$ is the set $\delta_i = \Delta p_i$, (i e the values obtained through natural selection). This is done by showing that the maximum of

$$2 \sum w_i \cdot \delta_i - \lambda \left\{ \sum \frac{\delta_i{}^2}{p_i} - \tfrac{1}{2}\sigma_A{}^2 \right\} - \mu \{\sum \delta_i - 1\},$$

where $\lambda$ and $\mu$ are Lagrange multiplies, is achieved when $\delta_i = \Delta p_i$. The proof is straightforward and is omitted.

This result and similar results have been extended and generalized in several ways. For details see Kimura (1958), Crow and Kimura (1956) and Wright (1949, 1955).

### 3.2. Two allele, non-random-mating populations

A considerable part of the preceding theory can be carried over to the case where random mating is no longer assumed. This generalization requires a return to the notation of Chapter 1, so that the frequencies of $A_1A_1$, $A_1A_2$, and $A_2A_2$ will be denoted $P$, $2Q$, $R$ respectively. While the equation $Q^2 = PR$ no longer necessarily holds, it is still true that $p = P + Q$, $q = Q + R$, and that the mean fitness of the population is

$$W = w_{11}P + 2w_{12}Q + w_{22}R. \tag{3.23}$$

For the random-mating case, $\Delta W$ is very close to the 'additive' component of the genetic variance, that is, to that component which can be removed by fitting a weighted least squares regression line to the fitnesses $w_{ij}$. It is therefore natural to seek a similar interpretation in the present more general case. The weighted sum of squares

$$P(w_{11} - W - 2\alpha)^2 + 2Q(w_{12} - W - \alpha - \beta) + R(w_{22} - W - 2\beta)^2$$

is minimized with respect to $\alpha$ and $\beta$ at those values $\alpha$ and $\beta$ satisfying the simultaneous equations

$$2P\alpha + Q(\alpha+\beta) = P(w_{11}-W) + Q(w_{12}-W), \qquad (3.24)$$

$$Q(\alpha+\beta) + 2R\beta = Q(w_{12}-W) + R(w_{22}-W). \qquad (3.25)$$

It is not necessary to solve these equations explicitly for $\alpha$ and $\beta$. On the other hand it is useful to note, firstly, that the solutions satisfy

$$p\alpha + q\beta = 0, \qquad (3.26)$$

and secondly that, if the definitions (3.6) and (3.7) are extended so that

$$E_1 = (Pw_{11} + Qw_{12} - pW)/p, \qquad (3.27)$$

$$E_2 = (Qw_{12} + Rw_{12} - qW)/q, \qquad (3.28)$$

then the sum of squares removed by the regression line is

$$\sigma_A{}^2 = 2p\alpha E_1 + 2q\beta E_2. \qquad (3.29)$$

Using (3.27) and (3.28), eqn. (3.29) may be written

$$\sigma_A{}^2 = 2pE_1(\alpha-\beta) = 2pq(E_1-E_2)(\alpha-\beta). \qquad (3.30)$$

The quantity $\alpha-\beta$ is called the *average effect* of $A_1$; for random mating populations this is identical to the average excess $E_1-E_2$, so that for such populations eqn. (3.30) reduce to eqn. (3.11). We now write

$$w_{11} = W + 2\alpha + d_{11},$$
$$w_{12} = W + \alpha + \beta + d_{12}, \qquad (3.31)$$
$$w_{22} = W + 2\beta + d_{22},$$

where the $d_{ij}$ are deviations of the true fitnesses $w_{ij}$ from the values obtained from the least-squares line.

To find $\Delta W$ it is necessary first of all to find $\Delta p$. The frequency of $A_1$ in the following generation is

$$p' = (w_{11}P + w_{12}Q)/W \qquad (3.32)$$

so that, using eqn. (3.27),

$$\Delta p = pE_1/W, \quad \Delta q = qE_2/W. \qquad (3.33)$$

It follows that

$$\Delta W = W' - W$$

$$= w_{11}P' + 2w_{12}Q' + w_{22}R' - W$$

$$= 2p'\alpha + 2q'\beta + d_{11}P' + 2d_{12}Q' + d_{22}R'$$

$$= 2p\alpha + 2q\beta + 2\alpha\Delta p + 2\beta\Delta q + d_{11}P' + 2d_{12}Q' + d_{22}R'.$$

Using eqns. (3.33) and (3.29), this may be written

$$\Delta W = \sigma_A{}^2/W + (d_{11}P' + 2d_{12}Q' + d_{22}R'). \qquad (3.34)$$

The definition (3.31) of the $d_{ij}$ implies that

$$d_{11}P + 2d_{12}Q + d_{22}R = 0,$$

so that the bracketed term on the right-hand side of eqn. (3.34) may be written

$$d_{11}\Delta P + 2d_{12}\Delta Q + d_{22}\Delta R. \qquad (3.35)$$

This is a weighted sum (with sum of weights zero) of the quantities $d_{ij}$, which will themselves often be small. Since also $W \approx 1$, it follows that to a close approximation

$$\Delta W = \sigma_A{}^2. \qquad (3.36)$$

If indeed $W = 1$, the condition that (3.36) holds exactly is that the expression (3.35) be zero. A sufficient condition for this is that each $d_{ij}$ be zero, but this condition is not necessary. Equation (3.31), taken with eqns. (3.24) and (3.25), shows that

$$d_{11}P + d_{12}Q = 0, \qquad (3.37)$$

$$Qd_{12} + Rd_{22} = 0. \qquad (3.38)$$

Thus $d_{12} = 0$ implies $d_{11} = d_{22} = 0$, in which case $\Delta W = \sigma_A{}^2$. If $d_{12} \neq 0$, then (3.35) can be written

$$\left\{ 2\frac{\Delta Q}{Q} - \frac{\Delta P}{P} - \frac{\Delta R}{R} \right\} Qd_{12}.$$

To a sufficiently close approximation, this will be zero when

$$2\frac{dQ}{Q} = \frac{dP}{P} + \frac{dR}{R}$$

or

$$Q^2 = \lambda PR, \qquad (3.39)$$

26

for any constant $\lambda$. This condition was given by Fisher (1941), and includes as a particular case the random mating population, for which $\lambda = 1$. Thus whenever eqn. (3.39) holds, the change in fitness must necessarily be positive; when (3.39) does not hold the argument following eqn. (3.34) suggests that fitness will usually, but not always, increase.

### 3.3. Discussion

The results of the previous sections indicate that the fitness of any population will usually increase, the rate of increase bearing a close relationship to the additive part of the genetic variance. Such a result has been called (Fisher, 1930) the 'Fundamental Theorem of Natural Selection', and has had considerable influence.

While this theorem is clearly of wide significance, the assumptions under which it has been derived must be kept in mind. Thus if the fitnesses themselves depend on genotypic frequencies, it is easy to devise situations in which the mean fitness does not increase, (Wright (1948), Moran (1962, p. 56)). For further examples see Fisher (1941) and Crosby (1949).

A more startling result is that when the fitness of any individual depends on his genetic constitution at two loci, (which is more realistic than the assumption that it depends on one), the Fundamental Theorem is no longer necessarily true. This occurs even if the two loci are unlinked. This anomalous behaviour will be examined in some detail in Section 8.3.

# CHAPTER 4

# Stochastic Treatment; Discrete Processes

So far we have ignored the chance fluctuations in gene frequency which must arise in a finite population. Clearly these should not be ignored entirely, and indeed one of our main aims is to discuss the extent to which the previous results have to be reconsidered in the light of such fluctuations. We shall attempt to do this by setting up a model to describe the behaviour of gene frequencies in a finite population.

## 4.1. Wright's model

As the simplest possible case, consider a diploid population having in each generation exactly $N$ individuals, so that there are $2N$ genes altogether at the locus in question. Suppose that in some way, to be chosen at our discretion, the population reproduces itself to form a daughter generation. Suppose also that once the daughter generation is formed, no further reproduction is possible for the parent generation, which may for mathematical purposes be considered as dying as soon as the daughter generation appears.

If there is no mutation, selection, or any other disturbing factor, then we know from Chapter 1 that gene frequencies tend to remain steady. It is therefore reasonable to assume in our model that if, in generation $t$, the number of $A_1$ genes is $X(t)$, then the number $X(t+1)$ of such genes in the following generation is a random variable with mean value $X(t)$. With this constraint, there are several possible models which could reasonably be devised; the most frequently used one, and the one we shall consider at some length, is due implicitly to Fisher (1930) and explicitly to Wright (1931). Under this model, $X(t+1)$ is a binomial variate with index $2N$ and parameter $X(t)/2N$. Explicitly, given that $X(t) = i$, the probability $p_{ij}$ that $X(t+1) = j$ is given by the equation

$$p_{ij} = \binom{2N}{j} \{i/2N\}^j \{1 - (i/2N)\}^{2N-j}. \tag{4.1}$$

Clearly $X(\cdot)$ is a Markovian variable with transition matrix $P = \{p_{ij}\}$.

It was remarked above that other models can be devised which also might reasonably describe the population behaviour. For a model in which the individuals in any given generation do not reproduce simultaneously, see Moran (1958). This model has several mathematical advantages over (4.1) and admits an extensive mathematical analysis (see, for example, Karlin and McGregor (1962)). However it is quite likely that (4.1) is more appropriate genetically, and since in any case we cannot expect any model to provide more than a rough guide to what occurs in nature, we shall concentrate attention in this book entirely on the model (4.1) and its various generalizations.

While, under the model (4.1), there is no tendency for any directed change in gene frequency, random sampling will ensure that eventually one of the two absorbing states, namely $X(\cdot) = 0$ and $X(\cdot) = 2N$, will be reached. Once this occurs, genetic variation is lost, and in view of the discussion in Chapter 3, it is of considerable interest to know how long such loss of variation will take.

There are two ways of attacking this problem, of which we consider firstly the more classical. If the eigenvalues of $P$ are

$$\lambda_i \ (i = 0, 1, 2, \ldots, 2N),$$

it is clear that, because there are two absorbing states, $\lambda_0 = \lambda_1 = 1$, and that if the remaining eigenvalues are distinct, $P^n$ can be written in the spectral form

$$P^n = C + \lambda_2^n P_2 + \ldots + \lambda_{2N}^n P_{2N}. \tag{4.2}$$

Here $C$ is a matrix having positive entries only in the first and last columns; these entries are absorption probabilities in $X(\cdot) = 0$ and $X(\cdot) = 2N$ respectively, for the various values of $X(0)$. If

$$|\lambda_2| > |\lambda_3| > \ldots > |\lambda_{2N}|,$$

then the rate at which $P^n$ converges to the limiting matrix $C$ is governed to a large extent by the value of $\lambda_2$. If $\lambda_2$ is very close to unity, this rate of approach will be slow and genetic variability will be lost slowly; if $\lambda_2$ is moderate or small, this rate will be quite rapid.

There are several methods for finding $\lambda_2$ by itself, but we shall consider initially a slightly more involved analysis which provides all eigenvalues of $P$ simultaneously.

Suppose that a non-singular matrix $Z$ and an upper triangular matrix $A$ can be found such that

$$PZ = ZA. \tag{4.3}$$

Since this equation implies $P = ZAZ^{-1}$, the eigenvalues of $P$ will be identical to those of $A$, which because of the special nature of $A$ are its diagonal elements. A matrix $Z$ for which an equation of the form (4.3) holds is

$$Z = \begin{pmatrix} 1 & 0 & 0 & 0 & 0 & 0 \\ 1 & 1 & 1^2 & 1^3 & & 1^{2N} \\ 1 & 2 & 2^2 & 2^3 & & 2^{2N} \\ 1 & 3 & 3^2 & 3^3 & \cdots\cdots & 3^{2N} \\ \vdots & \vdots & \vdots & \vdots & & \vdots \\ 1 & 2N & (2N)^2 & (2N)^3 & & (2N)^{2N} \end{pmatrix} \tag{4.4}$$

With this definition of $Z$, the $i$–$j$th element in $PZ$ is

$$\sum_l p_{il}\, l^j, \tag{4.5}$$

which can be written

$$E[\{X(t+1)\}^j | X(t) = i]. \tag{4.6}$$

Now the $i$–$j$th element of $ZA$ is of the form

$$\sum_{l=0}^{j} a_{lj}\, i^l. \tag{4.7}$$

Equations (4.6) and (4.7) taken together show that if, for each $j$, $(j = 0, 1, 2, \ldots, 2N)$, we can write

$$E[\{X(t+1)\}^j | X(t)] = a_{0j} + a_{1j}X(t) + a_{2j}\{X(t)\}^2 + \ldots + a_{jj}\{X(t)\}^j \tag{4.8}$$

then the $a_{jj}$ are the eigenvalues of $P$.

For small values of $j$, it is possible to check directly whether an expression of the form (4.8) can be obtained. This is obviously possible for $j = 0$, since we may choose $a_{00} = 1$. Similarly, since $E[X(t+1)|X(t)] = X(t)$, an equation of the form (4.8) is obtained by putting $a_{01} = 0, a_{11} = 1$. When $j = 2$, elementary binomial formulae show that

$$E[\{X(t+1)\}^2 | X(t)] = X(t) + \{1 - (2N)^{-1}\}\{X(t)\}^2,$$

which is of the form (4.8) with

$$a_{22} = 1 - (2N)^{-1}. \tag{4.9}$$

By using well known properties of the binomial distribution, it is found generally that an expression of the form (4.8) is possible for all $j$, and that

$$a_{jj} = 2N(2N-1)(2N-2)\ldots(2N-j+1)/(2N)^j,$$

$$(j = 1, 2, 3, \ldots, 2N). \tag{4.10}$$

Thus the largest non-unit eigenvalue of $P$ is $a_{22}$, and for large populations, this is extremely close to unity. We conclude that, while in the model (4.1) genetical variation must eventually be lost through random loss of one or other allele, the rate of such loss is extremely slow, so that genetic variation will be preserved for a long time. This conclusion, which will be reconsidered and supported from several different viewpoints later, may be thought of as the stochastic extension of the 'variation-preserving' portion of the Hardy-Weinberg theorem.

Before going any further, it is useful to consider a theorem which can be used to find $\lambda_2$ by itself, and which gives values quite readily of $\lambda_2$ in more complex situations.

*Theorem* 4.1. Let $Y(t)$ be a random variable which is zero for $X(t) = 0$, $X(t) = 2N$, and positive otherwise. If

$$E\{Y(t+1) | Y(t)\} = CY(t),$$

where $C$ is a constant, then $\lambda_2 = C$. The proof of this result follows directly from (4.2) and by noting that the above equation implies

$$E\{Y(t) | Y(0)\} = C^t Y(0)$$

for all $t$. Indeed the various values of $Y(\cdot)$, as $X(\cdot)$ assumes in turn the values $0, 1, 2, \ldots 2N$, will constitute the right eigenvector of $P$ corresponding to $\lambda_2$.

For the model (4.1), the simplest function satisfying the conditions required is

$$Y(t) = X(t)[2N - X(t)].$$

Clearly

$$E\{Y(t+1)\} = 2NE\{X(t+1)\} - E\{X(t+1)\}^2$$
$$= \{1 - (2N)^{-1}\}Y(t).$$

It follows directly from the theorem that $\lambda_2 = 1 - (2N)^{-1}$, as was found previously.

A second approach which shows that variation is lost slowly is to find the mean time to absorption at $X(\cdot) = 0$ or $X(\cdot) = 2N$, given the value of $X(0)$. Unfortunately, exact results seem to be impossible to find and consideration of this mean absorption time, which is a more easily interpreted and probably more relevant quantity than the leading eigenvalue, will be deferred to the next chapter.

### 4.2. Generalizations: the effective population size

We have seen in the previous section that even when no directed disturbing forces act, variation tends to be lost very slowly by random sampling (i.e. chance) effects. The largest non-unit eigenvalue of (4.1), namely $1 - (2N)^{-1}$, can be taken as a measure of this rate of dissipation of variation.

Suppose that in more general models, the largest non-unit eigenvalue (when no directed forces act) of the relevant transition matrix is

$$\lambda = 1 - (2N_e)^{-1}, \tag{4.11}$$

for some constant $N_e$. By analogy with eqn. (4.9) the constant $N_e$ will be called the 'effective population size' for the model under consideration.

Note that the effective population size so defined should not be attributed a significance beyond that deriving from its definition above. There are indeed other definitions of effective population size (Crow and Kimura (1963), Watterson (1964)) which for other purposes may be more relevant than the one just given.

To evaluate $N_e$ in specific cases, the following theorem is often useful.

*Theorem* 4.2. Let $X_1(\cdot), X_2(\cdot), \ldots, X_n(\cdot)$ be a set of jointly Markovian random variables with transition matrix

$$P = \{p_{i,j}\},$$

where

$$p_{i,j} = \text{Prob } \{X_1(t+1) = j_1, \ldots, X_n(t+1)$$

$$= j_n | X_1(t) = i_1, \ldots, X_n(t) = i_n\}.$$

Suppose that $X_i(\cdot) \geqq 0$ and $\sum_i X_i(\cdot) \leqq 2N$. Let $Y_1(\cdot), \ldots, Y_j(\cdot)$ be $r(r < (2N+1)^n)$ functions of the $X_i(\cdot)$ such that

$$E \begin{pmatrix} Y_1(t+1) \\ Y_2(t+1) \\ . \\ . \\ Y_r(t+1) \end{pmatrix} = M \begin{pmatrix} Y_1(t) \\ Y_2(t) \\ . \\ . \\ Y_r(t) \end{pmatrix}, \tag{4.12}$$

where $M$ is a matrix of constants and $E$ denotes expectation conditional on given values at generation $t$. Then the eigenvalues of $M$ are included amongst the eigenvalues of $P$. Further, if each $Y_i(\cdot)$ is zero for absorbing states of $P$, is non-negative otherwise and positive for at least one state of the vector $\{X_1(\cdot), \ldots, X_n(\cdot)\}$, then the largest non-unit eigenvalue of $M$ is identical to the largest non-unit eigenvalue of $P$.

We shall not prove this theorem other than to remark, firstly, that the proof follows from the spectral expansion (4.2), and secondly that the conditions imposed are stronger than are strictly necessary. The theorem will be used in the following sections to find the effective population size in various generalizations of the model considered in Section 4.1.

## 4.3. Bisexual populations

While the results of Section 1.3 have allowed us largely to ignore the bisexual nature of some populations, it is desirable to find the effective population size of such populations to obtain some idea of the effect of stochastic phenomena on their behaviour. Suppose that in each generation the total number of males and the total number of females are fixed at $N_1$, $N_2$ respectively, and that in generation $t$ the numbers of individuals of various genotypes at the locus '$A$' under consideration are

|  | $A_1A_1$ | $A_1A_2$ | $A_2A_2$ |  |
|---|---|---|---|---|
| Males | $k(t)$ | $N_1 - k(t) - l(t)$ | $l(t)$ | (4.13) |
| Females | $r(t)$ | $N_2 - r(t) - s(t)$ | $s(t)$ |  |

The probability that a gene chosen at random from the males of this generation is $A_1$ is

$$p_M(t) = \{N_1 + k(t) - l(t)\}/2N_1,$$

and the corresponding probability for females is

$$p_F(t) = \{N_2 + r(t) - s(t)\}/2N_2.$$

The probability that any individual, male or female, in generation $t+1$ is $A_1A_1$ is then $p_M(t)p_F(t)$, with similar probabilities for the other genotypes. The natural extension of (4.1) is that the probability that there are $kA_1A_1$ males, $lA_2A_2$ males, $rA_1A_1$ females and $sA_2A_2$ females in generation $t+1$, given the values (4.13), is

$$P\{k,l,r,s\} = \frac{N_1!N_2!}{k!\,(N_1-k-l)!\,l!\,r!\,(N_2-r-s)!\,s!}$$

$$\times \{p_M(t)p_F(t)\}^{k+r}\{(1-p_M(t))\,(1-p_F(t))\}^{l+s}$$

$$\times \{p_M(t)\,(1-p_F(t)) + p_F(t)\,(1-p_M(t))\}^{N_1+N_2-k-l-r-s}$$

(4.14)

Clearly the quantities $\{k(t), l(t), r(t), s(t)\}$ are jointly Markovian, and to find the leading eigenvalue of the matrix of which (4.14) is the typical element, it is necessary to find functions $Y_i(t)$ of $k(t)$, $l(t)$, $r(t)$ and $s(t)$ satisfying the conditions of Theorem 4.2. After some trial and error, it is found that the functions

$$Y_1(t) = 1 - N_1^{-2}\{k(t) - l(t)\}^2,$$

$$Y_2(t) = 1 - N_2^{-2}\{r(t) - s(t)\}^2,$$

$$Y_3(t) = 1 - N_1^{-1}N_2^{-1}\{k(t) - l(t)\}\{r(t) - s(t)\},$$

satisfy all the requirements mentioned, since they are zero in absorbing states, non-negative otherwise and positive for most states, and satisfy

$$EY_1(t+1) = \tfrac{1}{4}(1-N_1^{-1})\,Y_1(t) + \tfrac{1}{4}(1-N_1^{-1})\,Y_2(t) + \tfrac{1}{2}Y_3(t),$$

$$EY_2(t+1) = \tfrac{1}{4}(1-N_2^{-1})\,Y_1(t) + \tfrac{1}{4}(1-N_2^{-1})\,Y_2(t) + \tfrac{1}{2}Y_3(t),$$

$$EY_3(t+1) = \tfrac{1}{4}Y_1(t) + \tfrac{1}{4}Y_2(t) + \tfrac{1}{2}Y_3(t).$$

These equations show that the matrix $M$ (c.f. eqn. (4.12)) is

$$M = \begin{pmatrix} \tfrac{1}{4}(1-N_1^{-1}) & \tfrac{1}{4}(1-N_1^{-1}) & \tfrac{1}{2} \\ \tfrac{1}{4}(1-N_2^{-1}) & \tfrac{1}{4}(1-N_2^{-1}) & \tfrac{1}{2} \\ \tfrac{1}{4} & \tfrac{1}{4} & \tfrac{1}{2} \end{pmatrix}$$

One eigenvalue of $M$ is zero; the remaining two eigenvalues satisfy the equation

$$\lambda^2 - \{1 - \tfrac{1}{4}N_1^{-1} - \tfrac{1}{4}N_2^{-1}\}\lambda - \{\tfrac{1}{8}N_1^{-1} + \tfrac{1}{8}N_2^{-1}\} = 0.$$

The larger solution of this equation is

$$\lambda = \tfrac{1}{2}\{1 - \tfrac{1}{4}NN_1^{-1}N_2^{-1} + (1 - \tfrac{1}{16}N^2N_1^{-2}N_2^{-2})^{\frac{1}{2}}\},$$

where $N = N_1 + N_2$, and this is approximately

$$\lambda = 1 - \tfrac{1}{8}NN_1^{-1}N_2^{-1}. \tag{4.15}$$

Thus in accordance with eqn. (4.11) the effective population size is defined to be

$$N_e = 4N_1N_2/(N_1 + N_2).$$

Note that if $N_1 = N_2$, then $N_e = N$, as we might expect, while if $N_1$ is small and $N_2$ large, $N_e$ is approximately $4N_1$. Note also that when both $N_1$ and $N_2$ are large, $\lambda$ is very close to unity, so that loss of variation through random sampling is extremely slow. In such circumstances it is generally reasonable to treat the population as monoecious.

For sex-linked loci, taking males to be the heterogametic sex, similar methods lead to an effective population size of

$$N_e = 9N_1N_2(2N_1 + 4N_2)^{-1}.$$

Thus when $N_1 = N_2$, the effective population size is $\tfrac{3}{4}N$; again, when $N_1$ and $N_2$ are both large, loss of variation through random sampling is very slow.

## 4.4. Cyclic population size

Suppose that the population size assumes, in consecutive generations, the values $N_1, N_2, N_3, \ldots N_k, N_1, N_2, \ldots$. If the number of $A$ genes in generation $t$ is $X(t)$, and if the number of individuals in the population in this generation is $N_i$, then the natural extension of the preceding arguments shows that

$$EX(t+1)\{2N_{i+1}-X(t+1)\}/(2N_{i+1})^2$$
$$= \{1-(2N_{i+1})^{-1}\}X(t)\{2N_i-X(t)\}/(2N_i)^2. \quad (4.16)$$

Here all suffices of $N$ are taken modulo $k$. If the $N_i$ differ, there is no unique effective population size independent of $i$. However, iteration in eqn. (4.16) over $k$ consecutive generations shows that

$$EX(t+k)\{2N_i-X(t+k)\} = X(t)\{2N_i-X(t)\}\prod_{i=1}^{k}\{1-(2N_i)^{-1}\}, \quad (4.17)$$

so that it is reasonable to extend the definition of effective population size so that

$$\{1-(2N_e)^{-1}\}^k = \prod_{i=1}^{k}\{1-(2N_i)^{-1}\}.$$

To a close approximation this yields

$$kN_e^{-1} = N_1^{-1}+N_2^{-1}+ \ldots +N_k^{-1}. \quad (4.18)$$

Thus $N_e$ is very close to the harmonic mean of the $N_i$ and will therefore tend to be closer to the smaller $N_i$ values than to the larger. Thus if $k = 5$ and $N_i = 10^i (i = 1, 2, \ldots, 5)$, the effective population size is 45. For fluctuating population size there is therefore some scope for random loss of alleles, provided that the smallest size that the population ever assumes is quite small.

## 4.5. Geographical subdivision

No population will normally breed completely at random; amongst other things there must always be some effect of isolation by distance. Naturally, any mathematical model which attempts to describe at all faithfully the behaviour of populations where such isolation exists will probably prove to be impossible to analyse. Despite this, useful information can be found from several extremely simplified models, of which we consider now the following example.

Suppose that the total population, of size $N(H+1)$, is divided into $H+1$ subpopulations, each of size $N$, and that in each generation exactly $K$ genes, chosen at random from the $i^{\text{th}}$ subpopulation, enter the $j^{\text{th}}$ subpopulation, for all $i,j(i \neq j)$.

Suppose also that in generation $t$ there are $k_i(t)A_1$ genes in the $i^{\text{th}}$ subpopulation. If the model (4.1) applies in each subpopulation, the quantities $k_i(t)$ $(i = 1, 2, \ldots, H+1)$ will be jointly Markovian. To find the effective population size, it is necessary to find functions $Y_i(t)$ satisfying the conditions of Theorem 4.2; the form of the functions which have been found useful in this respect in previous sections leads us to try initially the function $Y_1(t)$ defined by

$$Y_1(t) = \sum_{i=1}^{H+1} k_i(t) [2N - k_i(t)].$$

After some algebra it is found that

$$EY_1(t+1) = (2N)^{-2}[4N^2 + H^2K^2 + K^2H - 2N - 4NKH] Y_1(t)$$
$$- (2N)^{-2}[K^2H + K^2 - 4KN] \sum_{i \neq j} \sum k_i(t) [2N - k_j(t)]. \tag{4.19}$$

This leads us to consider the function

$$Y_2(t) = \sum_{i \neq j} \sum k_i(t) [2N - k_j(t)].$$

It is found, again after considerable algebra, that

$$EY_2(t+1) = (2N)^{-2}[4HKN - K^2H^2 - K^2H] Y_1(t)$$
$$+ (2N)^{-2}[4N^2 + HK^2 + K^2 - 4NK] Y_2(t). \tag{4.20}$$

Since both $Y_1(t)$ and $Y_2(t)$ obey the conditions of Theorem 4.2, the effective population size can be found by considering the eigenvalues of the matrix

$$(2N)^{-2}\begin{pmatrix} 4N^2 + H^2K^2 + K^2H - 2N - 4NHK & 4KN - K^2H - K^2 \\ 4HKN - K^2H^2 - K^2H & 4N^2 + HK^2 + K^2 - 4HK \end{pmatrix}$$

If there were no subdivision in the population, the larger eigenvalue would be $1 - [2(H+1)N]^{-1}$. Thus if we denote the larger eigenvalue by $1 - \nu[2(H+1)N]^{-1}$, departures of $\nu$ from unity will provide a measure of the effect of geographical subdivision. Now $\nu$ satisfies the equation

$$\begin{vmatrix} \nu(H+1)^{-1} + H^2K^2(2N)^{-1} & 2K - K^2(H+1)(2N)^{-1} \\ + K^2H(2N)^{-1} - 1 - 2KH & \\ 2HK - K^2H(H+1)(2N)^{-1} & \nu(H+1)^{-1} + K^2(H+1)(2N)^{-1} - 2K \end{vmatrix} = 0$$

**37**

While this equation can be solved explicitly for $\nu$, a sufficiently close approximation will be obtained by ignoring terms of order $N^{-1}$, since $N$ is taken to be much larger than $H$ or $K$. The equation for $\nu$ then becomes

$$\begin{vmatrix} \nu - (H+1)(1+2HK) & 2K(H+1) \\ 2HK(H+1) & \nu - 2K(H+1) \end{vmatrix} = 0$$

or

$$\nu^2 - \nu(H+1)(1+2K+2KH) + 2K(H+1)^2 = 0. \qquad (4.21)$$

The relevant solution of this equation is

$$\nu = \tfrac{1}{2}(H+1)[1+2K+2HK$$
$$-\{1+4K^2+4K^2H^2-4K+4HK+8HK^2\}^{\frac{1}{2}}].$$

In order to obtain some idea of the effect of population subdivision on rate of loss of variation, values of $\nu$ for small values of $H$ and $K$ are indicated below in Table 4.1.

TABLE 4.1 Values of $\nu$ for various $H$ and $K$.

|   | | $H$ | | |
|----|--------|--------|--------|--------|
| $K$ | 1 | 2 | 4 | 8 |
| 1 | 0·8769 | 0·8953 | 0·9245 | 0·9527 |
| 2 | ·9378 | ·9460 | ·9612 | ·9752 |
| 4 | ·9688 | ·9726 | ·9803 | ·9878 |
| 8 | ·9844 | ·9862 | ·9901 | ·9939 |
| 16 | ·9922 | ·9930 | ·9950 | ·9970 |

This table has been taken directly from Moran (1962), to whom the above theory is due. Clearly, if there is only one migrant gene from any subpopulation to each of the other subpopulations, (i.e. $K = 1$) $\nu$ is quite close to unity. If $K$ exceeds unity there is negligible deviation of $\nu$ from unity. Thus while the model under consideration is highly artificial, it does seem to lead very forcibly to the conclusion that, even in more realistic models, the effective population size should be extremely insensitive to population subdivision. Thus, provided that no selective differences exist between the subpopulations, the collection of subpopulations can normally be regarded for practical purposes as a single large random-mating population. We shall confirm this result in a different way later, and consider subsequently what relevance this conclusion might have for evolutionary processes.

### 4.6. General offspring distributions

While the derivation of the model (4.1) is perfectly reasonable, it is possible to derive the model from a different, and sometimes more fruitful, point of view. Suppose that each gene leaves a random number of offspring genes according to some probability distribution $\{f_i\}$ and let $f(s) = \sum f_i s^i$. Suppose now that exactly $2N$ of the offspring genes from all the parent genes are chosen to make up the following generation, the remainder being discarded. We seek the probability $p_{ij}$ that, given $iA_1$ genes in the parent generation, there will be $jA_1$ genes in the daughter generation.

Now the generating function of the total number of genes produced is $[f(s)]^{2N}$, and the unconditional probability that exactly $2N$ genes are so produced is the coefficient of $s^{2N}$ in this expression. The unconditional probability that exactly $jA_1$ genes and $2N-jA_2$ genes are produced is the coefficient of $t^j s^{N2}$ in $[f(ts)]^i [f(s)]^{2N-i}$. It follows that for the conditional process

$$p_{ij} = \frac{\text{coefficient of } t^j s^{2N} \text{ in } [f(ts)]^i [f(s)]^{2N-i}}{\text{coefficient of } s^{2N} \text{ in } [f(s)]^{2N}}. \qquad (4.22)$$

Karlin and McGregor (1965) have shown that if $p_{ij}$ is defined in this way, the eigenvalues of the matrix $\{p_{ij}\}$ are

$$\lambda_0 = \lambda_1 = 1, \lambda_r = \frac{\text{coefficient of } s^{2N-r} \text{ in } [f(s)]^{2N-r}[f'(s)]^r}{\text{coefficient of } s^{2N} \text{ in } [f(s)]^{2N}}, \qquad (4.23)$$

$$(r = 2, 3, \ldots, 2N).$$

Clearly these eigenvalues depend on the form of $f(s)$. In particular, if we choose

$$f(s) = \exp \mu(s-1), \qquad (4.24)$$

then

$$p_{ij} = \frac{\text{coefficient of } t^j s^{2N} \text{ in } \exp \mu[i(ts-1)+(2N-i)(s-1)]}{\text{coefficient of } s^{2N} \text{ in } \exp \mu[2N(s-1)]}$$

$$= \binom{2N}{j} \left(\frac{i}{2N}\right)^j \left(\frac{2N-i}{2N}\right)^{2N-j}$$

which is eqn. (4.1). If the number of $A_1$ genes before the choice of the $2N$ genes for the next generation is referred to as the unconditional number of $A_1$ genes, and the number after such choice as the conditional number, the above shows that a sufficient condition for the probability distribution of the conditional number to have the distribution (4.1) is that the unconditional number have a Poisson distribution (with the same parameter for $A_1$ and $A_2$ genes); this condition is also necessary (Karlin and McGregor, (1964)). We check that with the Poisson distribution (4.22),

$$\lambda_r = \frac{\text{coefficient of } s^{2N-r} \text{ in } [\mu^r \exp 2N\mu(s-1)]}{\text{coefficient of } s^{2N} \text{ in } [\exp 2N\mu(s-1)]}$$

$$= (2N)!/\{(2N-r)!(2N)^r\}, \tag{4.25}$$

which agrees with (4.10).

As a second example, suppose that the unconditional distribution of daughter genes is binomial with generating function

$$f(s) = (q+ps)^c, (0 < p = 1 - q < 1, c \text{ an integer} \geq 2). \tag{4.26}$$

Then

$$p_{ij} = \frac{\binom{ci}{j}\binom{c(2N-i)}{2N-j}}{\binom{2Nc}{2N}},$$

and

$$\lambda_0 = \lambda_1 = 1, \lambda_r = \frac{c^r\binom{2Nc-r}{2N-r}}{\binom{2Nc}{2N}}, r = 2, 3, \ldots, 2N. \tag{4.27}$$

The largest non-unit eigenvalue is therefore

$$\lambda_2 = \frac{c^2\binom{2Nc-2}{2N-2}}{\binom{2Nc}{2N}}$$

$$\approx 1 - \frac{c-1}{c} \cdot \frac{1}{2N} \text{ for } N \text{ large.} \tag{4.28}$$

As a final example, if the unconditional generating function is

$$f(s) = q^\alpha(1-ps)^{-\alpha} \quad (0 < p = 1-q < 1, \alpha > 0),$$

then

$$p_{ij} = \frac{\binom{\alpha i + j - 1}{j}\binom{\alpha(2N-i)+2N-j-1}{2N-j}}{\binom{2N\alpha+2N-1}{2N}}, \qquad (4.29)$$

and

$$\lambda_0 = \lambda_1 = 1, \lambda_r = \alpha^r \frac{\binom{2N\alpha+2N-1}{2N-r}}{\binom{2N\alpha+2N-1}{2N}}, r = 2,3,\ldots,2N. \quad (4.30)$$

In particular,

$$\lambda_2 = \frac{2N\alpha-\alpha}{2N\alpha+1} \approx 1 - \frac{\alpha+1}{\alpha}\cdot\frac{1}{2N} \text{ for } N \text{ large.} \qquad (4.31)$$

It is possible to bring the results (4.25), (4.28) and (4.31) into a common form in the following way. If all unconditional distributions are normalized to have mean unity, then the variances of the unconditional distributions are $1$, $(c-1)/c$, $(\alpha+1)/\alpha$ respectively. Thus if this variance is denoted by $\sigma^2$, we can write in each case

$$\lambda_2 \approx 1 - (\sigma^2/2N). \qquad (4.32)$$

The remarkable fact is that (4.32) is a general result, so that generally

$$N_e = N/\sigma^2. \qquad (4.33)$$

We shall not prove this; indeed more useful results of this nature are available which take into account more effectively the diploid nature of the organism. Thus in a diploid population of $N$ individuals the effective population size is

$$N_e = (4N-2)/\{(\sigma^*)^2+2\}, \qquad (4.34)$$

where $(\sigma^*)^2$ is the variance of the distribution of the number of genes transmitted by each diploid individual. For discussion of this more useful result, see Fisher (1939), Haldane (1939), Wright (1939), Moran (1962, p. 94), Kimura and Crow (1963).

## 4.7. The asymptotic conditional distribution and the transient function

We have seen that if initially all genes in a population have moderate frequency, an extremely long time will usually pass before, as a result of random sampling only, one or other gene is lost from the population. Thus consideration of the transient behaviour of (4.1) is of particular interest, and in particular much attention has been paid in mathematical genetics to the following two questions: (a) what is the asymptotic ($t \to \infty$) distribution of the random variable $X(t)$, given $X(t) \neq 0$, $X(t) \neq 2N$; (b) what is the mean number of times that $X(\cdot)$ assumes the value $i$, given $X(0) = k$? The distribution (a) will be called the asymptotic conditional distribution, while the function (b) (which has $k$ as a parameter) will be called the transient function.

It is clear from the spectral expansion (4.2), coupled with the fact that

$$P_2 = x_2 y_2',$$

where $y_2$ and $x_2$ are the left and right eigenvectors of $P$ corresponding to $\lambda_2$, that the asymptotic conditional distribution is identical to the relevant elements in $y_2'$ (suitably normalized to add to unity). Unfortunately, the form of this eigenvector for the matrix (4.1) appears to be extremely complicated and does not admit a simple explicit formula. The most explicit formula available appears to be that given by Khazanie and McKean (1964), namely that if $\{b_i\}$ is the required distribution, then

$$b_j = \text{const } (2N)^{-j} \binom{2N}{j} \sum_{k=j}^{2N} (-1)^{k-j} \binom{2N-j}{k-j} u_k,$$

$$(j = 1, 2, \ldots 2N-1)$$

where

$$u_1 = 0, \ u_2 = 1,$$

$$u_n(\lambda_2 - \lambda_n) = a_{n2}\lambda_2 u_2 + a_{n3}\lambda_3 n_3 + \ldots + a_{n,n-1}\lambda_{n-1}u_{n-1} \ (n \geq 2),$$

with the $\lambda_i$ given by (4.25) and

$$a_{ij} = \sum_{l=0}^{j-1} \frac{(-1)^l (j-l)^{i-1}}{l!(j-l-1)!}, \ (j = 1, 2, \ldots, i).$$

Despite the complexity of this result, it will turn out that an extremely simple approximate formula for $\{b_i\}$ is available, and we will defer further consideration of this distribution until the next chapter, when this simple formula is derived.

Again, no exact formula has been found for the transient function. If the submatrix of $P$ corresponding to transitions between transient states is denoted by $Q$, then a formula suitable for numerical calculation for small $N$ is that the required function consists of the various elements in the $k^{th}$ row of $[I-Q]^{-1}$; the mean absorption time is the sum of these elements. Again, further consideration of this function will be deferred to the next chapter, where a simple approximate formula will be found.

### 4.8. Selection and mutation

So far in this chapter it has been assumed that no directed disturbing forces act in the population. This is a very unrealistic assumption and we most now consider the effect of such forces; in particular we must consider the effects of mutation and selection. Naturally, the previous results are not irrelevant as they should hold qualitatively for more complex situations as limiting cases for small selective pressures and zero mutation. Further, they are vitally important in showing that without disturbing forces, only very slow random changes in the population can be expected to occur; this will be important later since it explains why even very small selective forces have a major effect in determining the behaviour of the population.

Suppose that, among the $2N$ genes which form the mature zygotes in generation $t$, there are exactly $iA_1$ genes and thus $2N - iA_2$ genes. Then the expected frequencies of $A_1A_1$, $A_1A_2$ and $A_2A_2$ zygotes at the time of conception of generation $t+1$ are

$$(i/2N)^2, \ 2(i/2N)(1-i/2N), \ (1-i/2N)^2,$$

respectively. If these zygotes survive to reach maturity and in turn produce genes for the next generation in the ratios $1+s_1:1+s_2:1$, then the expected frequency of $A_1$ at the time of formation of the zygotes of generation $t+1$ is

$$\pi_i{}^* = \frac{(1+s_1)i^2+(1+s_2)i(2N-i)}{(1+s_1)i^2+2(1+s_2)i(2N-i)+(2N-i)^2}. \tag{4.35}$$

Further, if any such $A_1$ gene mutates to $A_2$ with probability $u$, and if any $A_2$ gene mutates to $A_1$ with probability $v$, then the expected frequency of $A_1$ at the formation of the zygotes of generation $t+1$ is

$$\pi_i = (1-u)\pi_i^* + v(1-\pi_i^*). \tag{4.36}$$

As an extension of (4.1) we now set up the following model. Suppose that, at maturity, there will be exactly $N$ individuals in any generation. The number of $A_1$ genes present in the individuals of generation $t+1$ is a random variable assuming one or other of the values $0, 1, 2, \ldots 2N$, and it is supposed that, given that the number of such genes in generation $t$ is $i$, the probability $p_{ij}$ that the number of such genes in generation $t+1$ is $j$ is given by

$$p_{ij} = \binom{2N}{j} \pi_i^j (1-\pi_i)^{2N-j} \tag{4.37}$$

$\pi_i$ being defined by eqns. (4.35) and (4.36).

Unfortunately it appears to be impossible to get any exact results whatsoever from this model, except that when $s_1 = s_2 = 0$ the eigenvalues of $P$ are

$$\lambda_r = (1-u-v)^r (2N)! / \{(2N-r)!(2N)^r\},$$
$$r = 0, 1, 2, \ldots, 2N. \tag{4.38}$$

It is again necessary to defer any quantitative consideration of this model until the following chapter. Qualitatively, however, we can say that when one or both of $u$ and $v$ are zero, eventual loss of one or other gene is certain. In such cases both the asymptotic conditional distribution and the transient function will exist. If both $u$ and $v$ are positive, neither this distribution nor this function is defined and interest centres on the stationary distribution of the frequency of $A_1$.

Although the exact form of this distribution is unknown, it is at least possible to find the lower moments easily, at least in some simple cases. Thus when $s_1 = s_2 = 0$,

$$E[X(t+1)|X(t)] = X(t)(1-u) + v[2N - X(t)]. \tag{4.39}$$

At equilibrium $E[X(t+1)|X(t)] = X(t)$, so that, using eqn. (4.39),

$$EX(\cdot) = 2Nv/(u+v). \tag{4.40}$$

Note that this is identical to the stable deterministic value found from eqn. (2.11). In a similar way it can be shown that

$$\text{Var } X(\cdot) = 4N^2 uv / \{(u+v)^2 (4Nu + 4Nv + 1)\}$$
$$+ \text{small order terms.} \tag{4.41}$$

A rough measure of likely variations in the number of $A_1$ genes from the mean (4.40) can be found by using Tchebycheff's inequality in conjunction with eqn. (4.41). The effect of population size is most easily seen if we consider the proportion $x(\cdot) = X(\cdot)/2N$ of $A_1$ genes, which from eqns. (4.40) and (4.41) has mean and variance

$$v/(u+v), \; uv/\{(u+v)^2(4Nu+4Nv+1)\} \qquad (4.42)$$

respectively. As a numerical example, suppose that $u = v = 10^{-6}$, and consider the three cases $N = 10^6$, $N = 10^7$, $N = 10^8$. Then by suitable choice of $a$ in each case, the Tchebycheff inequality

$$Pr\{|x - E(x)| > a\} \leq a^{-2} \, \text{Var} \, (x)$$

becomes

$$(N = 10^6) Pr\{|x - \tfrac{1}{2}| > 0 \cdot 75\} \leq 0 \cdot 05,$$

$$(N = 10^7) Pr\{|x - \tfrac{1}{2}| > 0 \cdot 25\} \leq 0 \cdot 05, \qquad (4.43)$$

$$(N = 10^8) Pr\{|x - \tfrac{1}{2}| > 0 \cdot 08\} \leq 0 \cdot 05.$$

The first inequality is, of course, useless, but the remaining two give some idea of the increasing tendency of $x(\cdot)$ to differ less and less from its mean value as the population size increases. We shall be able to make rather more precise estimates than those given above in the next chapter, when an approximate formula for the complete distribution of $x(\cdot)$ will be found.

CHAPTER 5

# Diffusion Approximations

## 5.1 The Fokker-Planck equation

It was noted in the previous chapter that many quantities and distributions of interest were exceedingly difficult to obtain for the model (4.1) and its generalizations. It is then perhaps surprising that extremely simple and easily derived approximations can be obtained for these quantities by using a diffusion approximation to the discrete process. Indeed, this approach has proved very powerful and useful in population genetics, and has provided a method for giving answers to a wide variety of questions, very few of which appear to admit an answer by any other method. The use of diffusion methods can be traced back to Fisher (1922); for particularly important uses see Wright (1931, 1945) and a review article by Kimura (1964).

While a rigorous derivation of diffusion formulae is possible, we shall be content here with a simple heuristic derivation and discussion. Consider a random variable diffusing over $[0, 1]$ in such a way that if, at any time $t$, the random variable assumes the value $x$, then the value $x + \delta x$ assumed at time $t + \delta t$ is a random variable such that

$$E(\delta x) = m(x)\delta t + 0(\delta t)^2, \tag{5.1}$$

$$\mathrm{Var}\,(\delta x) = v(x)\delta t + 0(\delta t)^2, \tag{5.2}$$

$$E(\delta x)^i = o(\delta t), \ i \geqq 3. \tag{5.3}$$

Here $m(x)$ and $v(x)$ are functions of $x$ but not of $t$, and are called respectively the drift and diffusion coefficients of the process. If $f(x;t)$ is the probability density of the random variable at time $t$, the theorem of total probability shows that

$$f(x;t+\delta t) = \int f(x - \delta x;t)\, g(\delta x;x - \delta x)\, d(\delta x), \tag{5.4}$$

where $g(\delta x; x - \delta x)$ is the probability density of a change in the value of the random variable from $x - \delta x$ to $x$ in the time interval $(t, t + \delta t)$.

Expansion of both $f(x-\delta x;t)$ and $g(\delta x;x-\delta x)$ in a Taylor series gives

$$f(x-\delta x;t) = f(x;t) - (\delta x)f'(x;t) + \tfrac{1}{2}(\delta x)^2 f''(x;t) + o(\delta x)^2, \quad (5.5)$$

$$g(\delta x;x-\delta x) = g(\delta x;x) - (\delta x)g'(\delta x;x)$$
$$+ \tfrac{1}{2}(\delta x)^2 g''(\delta x;x) + o(\delta x)^2. \quad (5.6)$$

Inserting these expressions in eqn. (5.4) and performing the integration, it appears that, provided the order of integration and differentiation can be interchanged,

$$f(x;t+\delta t) - f(x;t) = \delta t\left[ -\frac{\partial}{\partial x}\{m(x)f(x;t)\} \right.$$
$$\left. + \tfrac{1}{2}\frac{\partial^2}{\partial x^2}\{v(x)f(x;t)\}\right] + 0(\delta t)^2.$$

Dividing by $\delta t$ and letting $\delta t \to 0$, we derive the partial differential equation

$$\frac{\partial f(x;t)}{\partial t} = -\frac{\partial}{\partial x}\{m(x)f(x;t)\} + \tfrac{1}{2}\frac{\partial^2}{\partial x^2}\{v(x)f(x;t)\}, \quad (5.7)$$

for the density function of the random variable at time $t$. Equation (5.7) is known as the Fokker-Planck, or forward Kolmogorov, differential equation, and is of the greatest importance in the theory of population genetics.

Before considering solutions of eqn. (5.7), we shall show how a model such as (4.37) can be fitted into the framework required for diffusion methods. Since the random variable is to be restricted to the range $[0,1]$, we must first of all consider the frequency $x(t) = X(t)/2N$ of the gene $A_1$, rather than the absolute number $X(t)$ of $A_1$ genes. For the model (4.37), the expected change in $x$ between consecutive generations is

$$v + (1-u-v)\frac{(1+s_1)x^2 + (1+s_2)x(1-x)}{1+s_1x^2 + 2s_2x(1-x)} - x. \quad (5.8)$$

Assuming that $u$, $v$, $s_1$ and $s_2$ are $0(N^{-1})$ and are therefore small, this can be written as

$$x(1-x)\{s_1x + s_2(1-2x)\} - ux + v(1-x) + 0(N^{-2})$$
$$= x(1-x)\{\alpha_1 x + \alpha_2(1-2x)\}(2N)^{-1}$$
$$+ \{-\beta_1 x + \beta_2(1-x)\}(2N)^{-1} + 0(N^{-2}), \quad (5.9)$$

say, where $\alpha_1 = 2Ns_1$, $\alpha_2 = 2Ns_2$, $\beta_1 = 2Nu$ and $\beta_2 = 2Nv$. Under the same assumptions, the variance of the change in $x$ is

$$x(1-x)\,(2N)^{-1} + (0N^{-2}),\tag{5.10}$$

while higher moments are $0(N^{-2})$ or less. To fit the values (5.9) and (5.10) into the framework of (5.1) and (5.2), it is necessary to measure time in units of $2N$ generations. Having done this, while we cannot strictly let $\delta t \to 0$ for $N$ fixed, we expect that the distribution function of the random variable at time $t$ is very close to that whose density function is a solution of the equation

$$\frac{\partial f(x;t)}{\partial t} = \frac{-\partial}{\partial x}\,[\{x(1-x)\,(\alpha_1 x + \alpha_2(1-2x)) - \beta_1 x + \beta_2(1-x)\}f(x;t)]$$

$$+ \tfrac{1}{2}\,\frac{\partial^2}{\partial x^2}\,[x(1-x)f(x;t)].\tag{5.11}$$

The problem of how close solutions of eqn. (5.11), or various functions derived from (5.11), are to their exact (but unknown) values as defined by eqn. (4.37) is often rather difficult to answer. Some progress has been made in this direction in particular cases; see Moran (1960), Ewens (1964a, b, 1965) and Chia (1966). The general conclusion of these investigations seems to be that if the order of magnitude assumptions made above do in fact hold, the diffusion approximations to the model (4.37) are extremely accurate, and no doubt far more accurate than the model (4.37) itself is in describing the real world. When the order of magnitude approximations do not hold, diffusion methods are rather less accurate but are sometimes surprisingly good. Some examples will be given later which expand on these rather general remarks.

### 5.2. The drift function

If we formally integrate in eqn. (5.7) with respect to $x$, we find

$$\frac{\partial F(x;t)}{\partial t} = -m(x)f(x;t) + \tfrac{1}{2}\,\frac{\partial}{\partial x}\,\{v(x)f(x;t)\},\tag{5.12}$$

where $F(x;t)$ is the distribution function

$$\int_{0^-}^{x} f(y;t)\,dy.\tag{5.13}$$

This heuristic argument suggests the more strictly demonstrable fact (see Kimura, 1964)) that the rate at which, at time $t$, the probability that the random variable lies in $[0, x]$ is increasing is

$$-m(x)f(x;t) + \tfrac{1}{2} \frac{\partial}{\partial x} \{v(x)f(x;t)\}. \qquad (5.14)$$

We therefore call (5.14) the probability flux of the process; it is a most important quantity and will be used often in the subsequent analysis.

As an example of the use of (5.14), when $u$ and $v$ are both positive there will exist a stationary distribution for $x$ on $[0,1]$. Since a stationary distribution $f(x)$ has zero probability flux, $f(x)$ must satisfy

$$-m(x)f(x) + \tfrac{1}{2} \frac{d}{dx} \{v(x)f(x)\} = 0.$$

The solution of this equation is

$$f(x) = \text{const } [v(x)]^{-1} \exp \left[ 2 \int^{x} m(y)/v(y)\, dy \right], \qquad (5.15)$$

which is the general formula for the stationary distribution (c.f. Wright (1937, 1945)). Using the values for $m(x)$ and $v(x)$ derived from (5.9) and (5.10) we have

$$f(x) = \text{const } x^{2\beta_2 - 1}(1-x)^{2\beta_1 - 1} \exp \left[ 2\alpha_2 x + (\alpha_1 - 2\alpha_2) x^2 \right]. \qquad (5.16)$$

The general form of this curve bears out the comments made at the end of the previous chapter. If, for example, $u$ is extremely small, the curve approaches infinity at $x = 1$, indicating that it is quite likely that at any given time, the gene $A_1$ is fixed, or almost fixed, in the population. On the other hand, if $u$ and $v$ are not extremely small, (in comparison with $N^{-1}$), extreme values of the frequency of $A_1$ are unlikely. In the case $\alpha_1 = \alpha_2 = 0$, we are able to use eqn. (5.16) to give rather more accurate measures of likely variation from the population mean than are given by the relations (4.43). Thus, using eqn. (5.16), we have, when $u = v = 10^{-6}$,

$$(N = 10^6) \Pr\{|x - \tfrac{1}{2}| > 0 \cdot 184\} = 0 \cdot 05,$$

$$(N = 10^7) \Pr\{|x - \tfrac{1}{2}| > 0 \cdot 110\} = 0 \cdot 05,$$

$$(N = 10^8) \Pr\{|x - \tfrac{1}{2}| > 0 \cdot 035\} = 0 \cdot 05.$$

## 5.3. The asymptotic conditional distribution

When $s_1 = s_2 = u = v = 0$, eqn. (5.7) reduces to

$$\frac{\partial f(x;t)}{\partial t} = \frac{1}{2} \frac{\partial^2}{\partial x^2} \{x(1-x)f(x;t)\}. \tag{5.17}$$

Here a complete solution can be found, by separation of variables, in terms of an eigenfunction expansion. This has been done by Kimura (1955a), who obtains the expression

$$f(x;t) =$$
$$\sum_{i=1}^{\infty} \frac{4(2i+1)p(1-p)}{i(i+1)} T_{i-1}^1(1-2p) T_{i-1}^1(1-2x) \exp\{-\tfrac{1}{2}i(i+l)t\}. \tag{5.18}$$

In this expression, $p$ is the value of $x$ at time $t = 0$, and $T_{i-1}(x)$ is a Gegenbauer polynomial defined in terms of the hypergeometric function by

$$T_{i-1}^1(x) = \tfrac{1}{2}i(i+1)F(i+2, 1-i, 2, \tfrac{1}{2}(1-x)),$$

so that in particular

$$T_0{}^1(x) = 1, \ T_1{}^1(x) = 3x.$$

The expansion (5.18) is clearly the analogue of the spectral expansion (4.2), and remembering that unit time has been scaled to $2N$ generations, the eigenfunction $\exp\{-\tfrac{1}{2}i(i+1)t\}$ is clearly the analogue of $\lambda_{i+1}^{2Nt}$, namely

$$[\{(2N)(2N-1) \ldots \ldots (2N-i)/(2N)^{i+1}]^{2Nt}$$

$$\approx [1-(1+2+\ldots\ldots+i)/2N]^{2Nt}$$

$$\approx \exp\{-\tfrac{1}{2}i(i+1)t\}.$$

For large values of $t$, the asymptotic form of (5.18) is

$$f(x;t) = 6p(1-p) \exp(-t)$$

$$+30(1-2p)(1-2x)\exp(-3t)+\ldots\ldots \tag{5.19}$$

The factor $6p(1-p)$ is the analogue of the right eigenvector $\{k(2N-k)\}$ corresponding to $\lambda_2$. The leading term in (5.19) is independent of $x$, showing that for the diffusion process the asymptotic conditional distribution is

$$f(x) = 1, \ 0 < x < 1. \tag{5.20}$$

It is reasonable to infer that an approximately rectangular asymptotic conditional distribution will hold for the process (4.1); that such is indeed the case will be shown later (Section 6.6). This is certainly a far simpler expression than the exact (implicit) formula given in the previous chapter.

Eigenfunction expansions similar to (5.18) can be found in the more complicated models where more than two alleles are allowed (Kimura (1955b)), and also when selection and dominance are introduced. In the case of selection without dominance, (so that in the notation of (4.35) $s_1 = 2s_2 = 2s$ say) Kimura (1955c) has found that the leading eigenvalue $\exp(-t)$ in (5.19) should be replaced by

$$\exp[-t\{1 + (10)^{-1}s^2 - (7,000)^{-1}s^4 - (1,050,000)^{-1}s^6 \ldots \ldots \}]. \quad (5.21)$$

When dominance is allowed, the situation is rather more complicated. Kimura (1957) has found an expression analogous to (5.21) for the case of complete dominance. When heterozygotes have a selective advantage over both homozygotes, Robertson (1962) found numerically the interesting result that an increase in heterozygote advantage often decreases the value of the leading eigenvalue. This corresponds, in so far as the leading eigenvalue is important (and in the case under consideration it might be wrong to attach overmuch importance to it), to an increased rate of loss of variation as a joint result of random sampling and selection. This phenomenon, which is best explained intuitively, is most marked when the equilibrium gene frequency is near 0 or 1; clearly what happens is that for increased heterozygote advantage, the rate of approach through selection towards the equilibrium frequency is increased, while once this frequency is achieved random sampling becomes important and eventually leads to fixation. This result should be kept in mind when using the deterministic theory outlined in Section 2.1.

### 5.4. The backward differential equation

The (forward) differential equation (5.7) was derived by considering the process at the time points $0, t, t + \delta t$. An analogous equation can be found by considering the process at time points $0, \delta t, t$. If $f(x; p, t)$ denotes the probability density of the random variable $x$ at time $t$, given that $x = p$ at time zero, then

$$f(x; p, t + \delta t) = \int g(\delta p; p) f(x; p + \delta p, t) d(\delta p).$$

Expansion as in (5.5) and (5.6) shows that this equation reduces to the backward Kolmogorov equation

$$\frac{\partial f(x;p,t)}{\partial t} = m(p)\frac{\partial f(x;p,t)}{\partial p} + \tfrac{1}{2}v(p)\frac{\partial^2 f(x;p,t)}{\partial p^2}. \qquad (5.22)$$

Here an equally valid and indeed more useful equation is obtained by replacing $f(x;p,t)$ throughout by $F(x;p,t)$ to get

$$\frac{\partial F(x;p,t)}{\partial t} = m(p)\frac{\partial F(x;p,t)}{\partial p} + \tfrac{1}{2}v(p)\frac{\partial^2 F(x;p,t)}{\partial p^2}. \qquad (5.23)$$

By letting $x = 1-$, eqn. (5.23) shows directly that if $x = 1$ is an absorbing state, the probability $P_1(p;t)$ that $x$ has been absorbed at unity at or before time $t$ is a solution of the equation

$$\frac{\partial P_1(p;t)}{\partial t} = m(p)\frac{\partial P_1(p;t)}{\partial p} + \tfrac{1}{2}v(p)\frac{\partial^2 P_1(p;t)}{\partial p^2}. \qquad (5.24)$$

The probability $P_1(p)$ that $x$ is ever absorbed at unity is thus that solution of the equation

$$0 = m(p)\frac{dP_1(p)}{dp} + \tfrac{1}{2}v(p)\frac{d^2 P_1(p)}{dp^2} \qquad (5.25)$$

which satisfies the boundary condition $P_1(0) = 0$, $P_1(1) = 1$. Writing

$$\psi(x) = \exp\left[-2\int^x m(y)/v(y)dy\right], \qquad (5.26)$$

the required solution is clearly

$$P_1(p) = \left\{\int_0^p \psi(x)dx\right\} \Big/ \left\{\int_0^1 \psi(x)dx\right\}. \qquad (5.27)$$

The probability that absorption eventually occurs at $x = 0$ is similarly shown to be

$$P_0(p) = \left\{\int_p^1 \psi(x)dx\right\} \Big/ \left\{\int_0^1 \psi(x)dx\right\}. \qquad (5.28)$$

These quantities are naturally relevant only when $u = v = 0$. In this case, and also when only one of $u$ and $v$ is zero, considerable interest attaches also to the mean time until absorption occurs at one or other of these boundaries. This mean absorption time is an alternative measure of the rate of loss of variation through random sampling to that given by the eigenvalue method considered previously, and is indeed usually more useful and informative than the latter. We therefore turn to the problem of finding an equation for the mean absorption time which, since it depends (unlike the eigenvalues) on the initial frequency of $A_1$, will be denoted by $T(p)$.

If the probability density that absorption of $x$ either at zero or unity occurs at time $t$ is denoted by $\phi(p,t)$, the equation (5.23) yields

$$\frac{\partial \phi(p,t)}{\partial t} = m(p)\frac{\partial \phi(p,t)}{\partial p} + \tfrac{1}{2}v(p)\frac{\partial^2 \phi(p,t)}{\partial p^2}. \tag{5.29}$$

We use this equation to derive heuristically an equation for $T(p)$; for a more rigorous derivation, see Feller (1954). We have

$$-1 = -\int_0^\infty \phi(p,t)dt$$

$$= -\Big[t\phi(p,t)\Big]_0^\infty + \int_0^\infty t\,\frac{\partial \phi(p,t)}{\partial t}\,dt$$

$$= 0 + \int_0^\infty \left\{ m(p)\frac{\partial t\phi(p,t)}{\partial p} + \tfrac{1}{2}v(p)\frac{\partial^2 t\phi(p,t)}{\partial p^2} \right\} dt$$

$$= m(p)\frac{dT(p)}{dp} + \tfrac{1}{2}v(p)\frac{d^2 T(p)}{dp^2} \tag{5.30}$$

by interchange of the order of integration and differentiation. When both $x = 0$ and $x = 1$ are absorbing barriers, the solution of eqn. (5.30) is

$$T(p) = \int_0^1 t(x)dx, \tag{5.31}$$

where

$$t(x) = 2P_0(p)[v(x)\psi(x)]^{-1}\int_0^x \psi(y)dy,\ 0 \le x \le p, \tag{5.32}$$

$$t(x) = 2P_1(p)[v(x)\psi(x)]^{-1}\int_x^1 \psi(y)dy,\ p \le x \le 1. \tag{5.33}$$

53

The form of (5.31) is not fortuitous. It may be shown that the function $t(x)$ admits the interpretation that

$$\int_{x_1}^{x_2} t(x)dx \tag{5.34}$$

is the mean time that the random variable undergoing the diffusion process spends in the arbitrary interval $(x_1, x_2)$ before reaching one or other absorbing barrier. Thus the function $t(x)$ is the transient function of the process, for which we shall later find a number of uses.

For the moment, we note only that the transient function is to be distinguished clearly from the asymptotic conditional distribution discussed in the previous section; there is no similarity in interpretation or indeed of functional form between the two. It is also important to note that there is no similarity in interpretation or functional form between either of these two and the function derived by formally computing (5.15) when there is in fact no stationary distribution. These points are to be stressed, since in the literature some confusion has arisen by the assumption that these functions can be used interchangeably.

An instructive example of these differences occurs when there is no selection or mutation, so that the coefficients $m(x)$ and $v(x)$, defined by (5.1) and (5.2), become

$$m(x) = 0, v(x) = x(1-x).$$

We have seen that in this case the asymptotic conditional distribution is

$$a(x) = 1, 0 < x < 1. \tag{5.35}$$

Equations (5.32) and (5.33) show that the transient function is

$$t(x) = 2(1-p)/(1-x), \qquad 0 \leq x < p, \tag{5.36}$$

$$t(x) = 2p/x, \qquad\qquad p \leq x \leq 1, \tag{5.37}$$

while formal application of (5.15) yields

$$f(x) = \text{const } x^{-1}(1-x)^{-1}. \tag{5.38}$$

It is possible to interpret eqn. (5.38) as a limiting stationary distribution, approached but never achieved as mutation rates tend to zero, conditional on $x \neq 0$, $x \neq 1$. Clearly the functional forms of these three expressions are completely different.

Finally, it should be noted that the transient function is defined even if only one of $x=0$, $x = 1$ is absorbing. Taking $x = 0$ as the absorbing state we find (Ewens (1964b))

$$t(x) = 2[v(x)\psi(x)]^{-1} \int_0^x \psi(y)dy, \qquad 0 \le x \le p, \qquad (5.39)$$

$$t(x) = [v(x)\psi(x)]^{-1} \int_0^p \psi(y)dy, \qquad p \le x \le 1. \qquad (5.40)$$

Before moving on to applications of these formulae, it should be remarked in conclusion that our method of deriving the diffusion equation has assumed the existence of a single Markovian variate in terms of which the population behaviour may be described. Unfortunately there are several models (such as those considered in Sections 4.3 and 4.5) for which this is not the case, and for which a more complex treatment is required. It turns out that diffusion methods can be applied in such cases provided that certain regularity conditions are fulfilled. The form of these conditions, together with several applications, have been given by Watterson (1962, 1964); for reasons of simplicity we shall restrict ourselves in this book to cases where a single Markovian variate does in fact exist.

CHAPTER 6

# Applications

## 6.1. Absorption probabilities

Perhaps the most important conclusion to be obtained from the results of the previous chapter is that even a very small selective difference can make a striking difference to the fate of any gene. Indeed it will be shown that a selective difference which is impossible to observe experimentally (say 0·001 per cent) may be decisive in determining the progress of the frequencies of the various genes in a population.

To be specific, consider a locus admitting genotypes $A_1A_1$, $A_1A_2$ and $A_2A_2$ with fitnesses given by (2.6). If no mutation occurs, eqn. (5.27) may be used to find the probability that, for given values of $p$ (the initial frequency of $A_1$) and $N$ (the population size), the frequency of $A_1$ will eventually reach unity as a result of the joint effect of selection and random sampling.

As an example, suppose that $N = 10^6$, $p = \frac{1}{2}$, $s_1 = 0·00002$, $s_2 = 0·00001$. Equations (5.9) and (5.10) together show that

$$m(x) = 20x(1-x), \qquad v(x) = x(1-x).$$

From eqn. (5.27) it follows that

$$P_1(\tfrac{1}{2}) \approx 1 - \exp(-20) \approx 0·999999998. \qquad (6.1)$$

Thus even for a population as small as one million, a gene having the minute selective advantage of 0·00001 will, given initial frequency 0·5, almost certainly steadily increase in frequency to unity.

This is a startling result. It is perhaps even more startling if it is pointed out that while, when $p = \frac{1}{2}$, the mean increase per generation in the number of $A_1$ genes is only five, the probability that the number of $A_1$ genes actually decreases between consecutive generations is about 0·497.

The reason for the extremely strong effect of selective advantage is clear. While random fluctuations in the number of $A_1$ genes tend, over time, largely to cancel out, the selective advantage tends to provide a small but steady increase in numbers. Since we have seen that variation tends to be conserved for extremly long periods, sufficient time will elapse for this small but steady increase to dominate the effect of random fluctuations.

While it would be wrong to attribute overmuch importance to a single numerical result, the value (6.1) is in fact typical of what occurs in a very wide range of cases. Further, while much of the original work in population genetic assumed as tentative estimates selective differences of the order of 1 per cent, recent experiments seem to suggest that selective differences are often of the order of 5–10 per cent or even more. In such cases, unless the population size is extremely small and $p$ is very close to zero, the same behaviour as that noted in the numerical example above will occur.

On the other hand, it should be noted that several of the results of Chapter 4 are relevant to this argument. Firstly, the expression 'population size' refers to that generation of the population currently breeding; in humans this may be a third or a quarter of the total population. Further, markedly unequal numbers in the two sexes will increase the importance of the stochastic factor, as will marked cyclic fluctuations in population size. On the other hand, non-Poisson offspring distribution (with a larger variance than mean) would probably not alter the effective population size by a large factor, while geographical sub-division probably has a very small effect.

The conclusions derived above do not apply if $p$ is extremely small. In particular, if $p = (2N)^{-1}$ (so that $A_1$ is a new mutant in an otherwise pure $A_2A_2$ population), then (5.27) yields

$$P_1\{(2N)^{-1} \approx (2N)^{-1} \bigg/ \int_0^1 \psi(x)dx. \tag{6.2}$$

When $s_1 = 2s_2$, this may be written

$$P_1\{(2N)^{-1} \approx 2s_2/\{1 - \exp(-4Ns_2)\}.$$

If $s_2 > N^{-1}$, this becomes

$$P_1\{(2N)^{-1}\} \approx 2s_2. \tag{6.3}$$

Note that this is independent of $N$. The reason for this is that once the number of mutants exceeds about $3/s_2$ there is negligible chance that, by random sampling, the mutant will die out, no matter how large the population size may be. Indeed, eqn. (6.3) will be re-obtained later using branching processes for which no finite limit is placed on the population size.

Another special case of (5.27) occurs when $s_2 = 0$. Here

$$\psi(x) = \exp(-\alpha_1 x^2)$$

so that

$$P_1\{(2N)^{-1}\} = (2N)^{-1} \Big/ \int_0^1 \exp(-\alpha_1 x^2)dx \qquad (6.4)$$

$$= (2\alpha_1)^{\frac{1}{2}}(2N)^{-1} \Big/ \int_0^{(2\alpha_1)^{1/2}} \exp(-\tfrac{1}{2}y^2)dy. \qquad (6.5)$$

If $(2\alpha_1)^{\frac{1}{2}} > 3$, the upper terminal in the integral may effectively be replaced by infinity, in which case (6.5) becomes

$$P_1\{(2N^{-1})\} \approx (2s_1/\pi N)^{\frac{1}{2}}. \qquad (6.6)$$

This result is due to Kimura (1957); similar results were obtained previously by Wright (1952) and Haldane (1927b).

## 6.2. Mean absorption times

We turn now to the problem of the time required for absorption at $x = 0$ or $x = 1$. The most interesting case is where $s_1 = s_2 = 0$, for which the transient function is given by eqns. (5.36) and (5.37). It follows that the mean absorption time $T(p)$ is (in units of $2N$ generations)

$$T(p) = -2\{p\ln p + (1-p)\ln(1-p)\}. \qquad (6.7)$$

Thus for example when $p = \frac{1}{2}$, the mean time until one or other gene is lost as a result of random sampling alone is about $2 \cdot 8N$ generations; this will be an extremely long time for most populations. Naturally this result is intimately connected with the value of the eigenvalue (4.9), and clearly like that eigenvalue would require some adjustment for the more general models of Section 4.2. Such adjustment would usually be identical to that made for the eigenvalue, as the following demonstration shows.

Because the concept of effective population size has been used above to consider how quickly, in various circumstances, random sampling acting alone leads to loss of one or other gene, it is most relevant to us when no directed forces are operating. When this is the case, it is often possible to find a variable $x(t)$ (usually the frequency of a given gene in generation $t$) with the properties

$$0 \leqq x(t) \leqq 1, \quad E\{x(t+1)|x(t)\} = x(t), \tag{6.8}$$

$$Ex(t+1)\{1-x(t+1)\} = \{1-(2M)^{-1}\}x(t)\{1-x(t)\}. \tag{6.9}$$

If such a variable exists, it follows immediately from Theorem 4.1 that the effective population size is $M$. Further, the condition (6.8) and (6.9) imply

$$\text{Var } \{x(t+1)-x(t)\} = (2M)^{-1}x(t)\{1-x(t)\}. \tag{6.10}$$

If the random variable $x(t)$ can be treated as a diffusion variate, it follows that its drift and diffusion coefficients are

$$m(x) = 0, \quad v(x) = (N/M)x(1-x)$$

respectively, so that the mean time $T(p)$ until $x(\cdot)$ reaches either zero or unity is

$$T(p) = -2(M/N)\{p\,lnp + (1-p)ln(1-p)\}.$$

Comparison of this value with (4.9) and (4.11) shows that, provided the above assumptions can be met, any alteration in the effective population size for the generalized models of Chapter 4 will be mirrored by a corresponding alteration to $T(p)$.

In any event, it is clear that if $p$ is not extremely close to 0 or 1, a long time will usually pass before one or other gene is lost as a result of random sampling; using eqn. (6.7), this mean time exceeds $N$ generations so long as $0 \cdot 07 < p < 0 \cdot 93$.

Particular interest attaches to the case $p = (2N)^{-1}$. Here

$$T\{(2N)^{-1}\} \approx N^{-1}\{1+ln2N\}, \tag{6.11}$$

$$t(x) = 2/x, \quad (2N)^{-1} \leqq x < 1. \tag{6.12}$$

Converting to generations, the mean number of generations until one or other gene is lost is about

$$2+2\,ln\,2N, \tag{6.13}$$

while the mean number of such generations for which the number of $A_1$ genes is $i$ is approximately

$$2/i, \quad (i = 1, 2, \ldots . 2N-1) \tag{6.14}$$

These results cannot both be exactly correct, since the sum of the terms in (6.14) is not quite equal to (6.13), but they are sufficiently accurate for any reasonable purpose.

When selective differences are present, the formula for $T(p)$ is by no means as simple as (6.7). Indeed we shall see that the mean time until one or other gene is lost will depend largely on the fitnesses involved and will to some extent be independent of population size. (We have seen in eqn. (6.3) that a similar phenomenon can occur for survival probabilities.) This is illustrated by two examples. When $s_1 = 2s_2$, eqns. (5.9), (5.10), (5.27), (5.32) and (5.33) show that $t(x)$ assumes the form

$$t(x) = \{1 - P_1(p)\}\{\alpha_2 x(1-x)\}^{-1}\{\exp(2\alpha_2 x) - 1\}, \; 0 \leq x \leq p, \quad (6.15)$$

$$t(x) = P_1(p)\{\alpha_2 x(1-x)\}^{-1}\{1 - \exp - 2\alpha_2(1-x)\}, \; p \leq x \leq 1, \quad (6.16)$$

where $\alpha_2 = 2Ns_2$ and

$$P_1(p) = \{1 - \exp(-2\alpha_2 p)\}/\{1 - \exp(-2\alpha_2)\}.$$

Sometimes the fitnesses of $A_1 A_1$ and $A_1 A_2$, and the initial frequency of $A_1$, are jointly sufficiently large to make eventual fixation of $A_1$ effectively certain; the expression for $P_1(p)$ shows that this will be the case, approximately, when $\alpha_2 p > 3$. It follows from eqns. (6.15) and (6.16) that the frequency $x$ of $A_1$ will be less than its initial frequency $p$ only for a negligible time, while for $p \leq x_1 < x_2 \leq 1$, the mean time that $x$ spends in the range $(x_1, x_2)$ is

$$t(x_1, x_2) = \int_{x_1}^{x_2} [\alpha_2 x(1-x)]^{-1}[1 - \exp\{-2\alpha_2(1-x)\}]dx. \quad (6.17)$$

If the time is measured in terms of generations rather than in units of $2N$ generations, this time becomes

$$t(x_1, x_2) = \int_{x_1}^{x_2} [s_2 x(1-x)]^{-1}[1 - \exp\{-2\alpha_2(1-x)\}]dx. \quad (6.18)$$

In particular, the mean number of generations until $A_1$ is fixed is

$$T(p) = \int_{p}^{1} [s_2 x(1-x)]^{-1}[1 - \exp\{-2\alpha_2(1-x)\}]dx. \quad (6.19)$$

Equation (6.18) is very useful in demonstrating the extent to which

60

stochastic elements are important in the gene fixation process being considered. Under the deterministic treatment of Section 2.1, the mean number of generations (c.f. eqn. (2.9)) spent by $x$ in the range $(x_1, x_2)$ is

$$\int_{x_1}^{x_2} [s_2 x(1-x)]^{-1} dx. \qquad (6.20)$$

This will agree with eqn. (6.18) provided that the term

$$\exp \left\{ -2\alpha_2 (1-x) \right\}$$

is negligible. This is so for $x \leqq 1 - 3\alpha^{-1}$ (approximately), so that for values of $x$ up to this value, the time spent in any frequency range is independent of the population size and is controlled completely by selective pressures.

However, as $x$ approaches unity, the formula (6.20) is misleading in that it suggests that extremely long times are required for extremely small increases in frequency, whereas eqn. (6.18) shows that when stochastic fluctuations are taken into account, such is not the case. The extent to which this stochastic effect is important is illustrated numerically in Table 6.1.

TABLE 6.1. Approximate times spent in various frequency ranges under deterministic and stochastic treatments, calculated from (6.18) and (6.20).

$s_1 = 0.001$, $s_2 = 0.0005$, $N = 10^6$, $p = 0.1$.

| Range | 0·1–0·99 | 0·99–0·999 | 0·999–0·9999 | 0·9999–0·99999 | 0·99999–0·999999 |
|---|---|---|---|---|---|
| Deterministic | 13,600 | 4,600 | 4,600 | 4,600 | 4,600 |
| Stochastic | 13,600 | 4,500 | 2,300 | 350 | 40 |

The values in this table verify the mathematical predictions; in particular they confirm that there is practically no difference in the times spent in various ranges calculated by the two formulae for values of $x$ less than $1 - 3\alpha_2^{-1} = 0.997$.

Secondly, we shall consider the important case $s_1 = s_2$ (i.e. $A_1$ dominant to $A_2$). Under a deterministic treatment, the number of generations spent in any frequency range $(x_1, x_2)$ is (c.f. eqn. (2.8))

$$\int_{x_1}^{x_2} [s_1 x(1-x)^2]^{-1} dx. \qquad (6.21)$$

If $s_1$ and $p$ are together sufficiently large to allow the approximation $P_1(p) = 1$ to be used, the time spent in the same frequency range under a stochastic treatment (c.f. eqn. (5.33)) is

$$\int_{x_1}^{x_2} g(x)[s_1 x(1-x)^2]^{-1}dx, \qquad (6.22)$$

where

$$g(x) = 2\alpha_1^{\frac{1}{2}}(1-x) \exp\left\{-\alpha_1(1-x)^2\right\} \int_0^{\alpha_1^{1/2}(1-x)} \exp z^2 dz. \qquad (6.23)$$

Now it is not difficult to show that the function $h(y)$, defined by

$$h(y) = 2y \exp\left(-y^2\right) \int_0^y \exp z^2 dz,$$

is asymptotically unity for large $y$, and that, for $y > 5$, $h(y)$ differs from unity only by a negligible amount. Thus for $1-x > 5\alpha_1^{-\frac{1}{2}}$, we can take $g(x) = 1$, so that the times indicated by the 'deterministic' eqn. (6.21) and the 'stochastic' eqn. (6.22) will agree for this range. To this extent, the time required for fixation of the favoured gene is independent of population size and is determined solely by selective forces.

For values of $x$ larger than $1-5\alpha_1^{-\frac{1}{2}}$, $g(x)$ is of order $(1-x)^2$ and there will be considerable difference between the times calculated under deterministic and stochastic treatments. Indeed this difference will be much larger than that in the previous example, where the correction factor $g(x)$ was only of order $(1-x)$ near $x = 1$. This is illustrated in Table 6.2 below.

There is little point in giving further numerical examples; it is clear that the time spent in various frequency ranges is determined entirely

TABLE 6.2. Approximate times spent in various frequency ranges under deterministic and stochastic treatments, calculated from (6.21) and (6.22).

$s_1 = s_2 = 0.001$, $N = 10^6$, $p = 1$.

| Range | 0·1–0·8 | 0·8–0·95 | 0·95–0·98 | 0·98–0·99 | 0·99–0·999 | 0·999–0·99999 |
|---|---|---|---|---|---|---|
| Deterministic | 7,500 | 21,800 | 25,700 | 50,700 | 902,300 | 99,000,000 |
| Stochastic | 7,500 | 23,400 | 24,300 | 30,300 | 34,500 | 4,000 |

by selective forces until a point is reached when the frequency of $A_1$ is sufficiently close to unity, where stochastic effects become important. The effect of stochastic fluctuations is that the favoured gene should become fixed rather more quickly than deterministic theory would suggest.

We conclude this section by mentioning that in the event that reverse mutation $A_1 \rightarrow A_2$ (at rate $u$) exists, we should be concerned not so much with the time required for the frequency of $A_1$ to reach unity as with the time required for the frequency of $A_1$ to reach the selection-mutation equilibrium point $x_0$. When $s_1 = 2s_2$, this point is (c.f. eqn. (2.16))

$$x_0 = 1 - u/s_2. \qquad (6.24)$$

To examine processes involving mutation, extra terms should strictly be included in all the formulae; however, because mutation rates are so small, omission of such terms should cause little loss of accuracy. Comparison of (6.18) and (6.20) shows that use of the deterministic model should be accurate to within 1 per cent provided that for all $x \leqq x_0$,

$$1 - \exp -2\alpha_2(1-x) \geqq 0.99.$$

This will certainly be so if

$$1 - \exp -2\alpha_2(1-x_0) \geqq 0.99.$$

Inserting the expression (6.24), this requirement reduces to

$$N \geqq 1 \cdot 15u^{-1}. \qquad (6.25)$$

Note that this condition is independent of $s_2$. If values of $u$ between $10^{-5}$ and $10^{-6}$ may be taken as typical, then deterministic theory will be adequate for population sizes exceeding about $10^5$ or $10^6$. For the case $s_1 = s_2$, the same result is found; when $s_2 = 0$ the requirement is that the population size should exceed about $10^6$ or $10^7$. If a 'large' population is defined as one in which random fluctuations can be ignored, then so far as the facet of population behaviour under consideration is concerned, a population is becoming 'large' once its size exceeds one million. This same question will be considered from other points of view later (with much the same conclusion); for a different set of considerations again leading to the same result, see Haldane and Jayakar (1963).

63

### 6.3. Stationary distributions

When the mutation rates $u(A_1{\to}A_2)$ and $v(A_2{\to}A_1)$ are positive, a stationary distribution for the frequency $x$ of $A_1$ over $[0,1]$ may be defined. We now examine some of the consequences of the expression (5.16) for this distribution.

Suppose firstly that no selective forces act. In this case, eqn. (4.38) shows that the largest non-unit eigenvalue of the matrix (4.37) is

$$\lambda_1 = 1-u-v. \qquad (6.26)$$

Since $u$ and $v$ are usually of order $10^{-5}$ at most, it is clear that $\lambda_1$ is very close to unity. Thus, starting from arbitrary frequencies, an exceedingly long time (of order $10^5$ generations) may be required until the form of the stationary distribution has been reached, so that any formula for stationary distributions can only be used when environmental conditions are sufficiently stable for this to occur. (When selection also operates, a considerably shorter time than this will usually suffice.)

When $s_1 = s_2 = 0$, the diffusion approximation (5.16) to the stationary distribution of the frequency $x$ of $A_1$ is

$$f(x) = \text{const } x^{2\beta_2-1}(1-x)^{2\beta_1-1}, \qquad 0\le x\le 1, \qquad (6.27)$$

where $\beta_1 = 2Nu, \beta_2 = 2Nv$. Since (6.27) is a beta distribution, it may be shown immediately from well known properties of such distributions that

$$E(x) = \beta_2/(\beta_1+\beta_2), \text{ Var } (x) = \beta_1\beta_2/(\beta_1+\beta_2)^2(2\beta_1+2\beta_2+1). \quad (6.28)$$

Supposing for convenience that $u = v = 5\times 10^{-6}$, it is interesting to consider the form of (6.27), together with the mean and variance (6.28), for representative values of $N$. Clearly for all $N$, $E(x) = \frac{1}{2}$. When $N < \frac{1}{2}\times 10^5$, the distribution of $x$ is U-shaped. Thus when $N \lll \frac{1}{2}\times 10^5$, the most likely situation, at any one time, is that where one or other gene is quite rare or even absent. For $N = \frac{1}{2}\times 10^5$ the distribution is rectangular, so that all values in $[0,1]$ are equally likely. For $N > \frac{1}{2}\times 10^5$ the distribution is unimodel; when $N = 10^6$ the distribution is closely concentrated around $x = \frac{1}{2}$, the standard deviation being $0\cdot 078$. In this case the frequency of $A_1$ is unlikely, at any time, to differ from $\frac{1}{2}$ by more than about $0\cdot 16$. For $N = 10^7$ the standard deviation is $0\cdot 025$. Thus if mutation at rate $5\times 10^{-6}$ can be taken as typical, populations of size one million or more will not usually exhibit any great deviation from the mean frequencies of the two genes. Since this mean frequency is identical to the deterministic equilibrium frequency, it is clear that for populations of this size we may expect a quasi-deterministic behaviour. This agrees with the result found in the previous section concerning the size of a 'large' population.

Two different sorts of behaviour are to be expected when selective differences exist, the first when $|s_1| > |s_2|$ and the second when $|s_1| < |s_2|$. In the first case, the frequency $x$ will usually be very close to zero $(s_1 < 0)$ or unity $(s_1 > 0)$. These two cases are effectively mirror-images and here only the case $s_1 < 0$ will be considered in any detail. It will be assumed that $s_1$ and $s_2$ are at least of order $10^{-4}$; this is probably the most relevant range for natural populations.

In the case $s_1 < s_2 < 0$, since $x$ will usually be very small, it is reasonable to ignore terms of order $x^2$ in (5.16) together with the term $(1-x)^{2\beta_1-1}$, to find approximately

$$f(x) = \text{const } x^{2\beta_1-1} \exp 2\alpha_2 x, \qquad 0 \le x \le 1. \qquad (6.29)$$

While an exact determination of the constant is possible, a sufficiently close approximation should be obtained, when $\alpha_2 < -3$, by assuming that $x$ may take values in $(0,\infty)$ rather than $(0,1)$. This assumption gives the value $|2\alpha_2|^{2\beta_2}\Gamma(2\beta_2)$ for the constant; it follows that the mean and variance of the distribution of $x$ are

$$\beta_2/|\alpha_2|, \qquad \beta_2/(2|\alpha_2|^2), \qquad (6.30)$$

respectively. The mean value coincides with the value (2.17) obtained by using a deterministic treatment, while the variance now provides a measure of likely deviations from this value in the stochastic case. We defer for the moment further consideration of these results, since we shall re-derive them later by a rather different method.

In the case $s_2 > s_1 > 0$, deterministic theory leads, in the absence of mutation, to an internal stable equilibrium point for $x$. In the stochastic case, in the absence of mutation, such a point can only be called quasi-stable; eventually $x$ must reach zero or unity, after which no further fluctuation can occur. Normally this will take an extremely long time, and since in any event we can suppose that mutation does occur, it is useful to consider the form of the stationary distribution applicable when $u, v > 0$.

65

For the deterministic treatment (Section 2.3), the presence of mutation causes only a minor disturbance from the equilibrium value

$$x = s_2/(2s_2 - s_1). \tag{6.31}$$

We therefore expect that the mean value of the distribution (5.16) does not differ appreciably from this value. Unfortunately, the complexity of (5.16) makes it difficult to derive an exact value for the mean. As a guide to the sort of result that can be expected, it is useful to consider the numerical case $N = \frac{1}{4} \times 10^5$, $u = v = 5 \times 10^{-6}$, $s_1 = 0$, $s_2 = 2 \times 10^{-3}$. For these values, we find $\beta_1 = \beta_2 = \frac{1}{4}$, $\alpha_1 = 0$, $\alpha_2 = 100$, and

$$f(x) = Cx^{-\frac{3}{4}}(1-x)^{-\frac{3}{4}} \exp 200x(1-x), \tag{6.32}$$

where $C$ is a normalizing constant. By symmetry, the mean value of $x$ is $\frac{1}{2}$. An indication of the likely fluctuations from this mean can be found by computing the variance of $x$. While this can be done exactly (c.f. Whittaker and Watson, 1962, p. 353, ex. 1), the method used involves summation of a very slowly converging series, and appears to be impracticable in the present context. As a reasonable alternative, possibly the best way of getting some idea of the values that $x$ is most likely to assume is to compare the integral of $f(x)$ over various subranges. Thus, for example, the probability that $x$ assumes a value in $(0, 10^{-4})$ is approximately

$$C \int_{0}^{10^{-4}} x^{-\frac{3}{4}} dx = 0 \cdot 02\, C, \tag{6.33}$$

while the probability that $x$ assumes a value in $(\frac{1}{2} - \frac{1}{2} \times 10^{-4}, \frac{1}{2} + \frac{1}{2} \times 10^{-4})$ is approximately

$$0 \cdot 0002 \times C \times \exp (50). \tag{6.34}$$

This is about $10^{20}$ times larger than (6.33). Similar comparisons over other frequency ranges give analogous results, so that it appears that despite the fact that $f(x)$ is unbounded near $x = 0$, $x = 1$, large fluctuations of $x$ from $\frac{1}{2}$ are extremely unlikely. This implies that, in the case under consideration, selective forces are strong enough to make stochastic fluctuations negligible.

Naturally a less extreme result will hold if selective forces are weaker. Thus if $s_2$ decreased to $2 \times 10^{-4}$, while the other parameters remain unchanged, (6.33) remains unchanged while (6.34) is decreased to $0 \cdot 0002 \times C \times \exp 5$, which is of the same order of magnitude as (6.33).

## 6.4. The maintenance of alleles by mutation

In this section we consider the following model: at a certain locus 'A', alleles from the infinite sequence $A_1, A_2, \ldots \ldots$ can occur. There are no selective differences between the different alleles, and at any generations any gene present in the population will mutate, with probability $u$, to form an entirely new allele which does not otherwise exist, and has never previously existed, in the population. We ask what is the mean number, at equilibrium, of different alleles present in the population? Questions of this sort are posed by the study of the number and extent of selectively equivalent isoalleles, considered in particular by Lewontin and Hubby (1966).

In answering this question, we shall suppose that the population is of fixed size $N$, and that if the number of $A_k$ genes in any generation is $i$, then the probability that the number of such genes in the following generation is $j$ is given by eqn. (4.37), where now

$$\pi_i = i(1-u)/2N. \tag{6.35}$$

Together with the problem under consideration, we may ask an associated question, namely 'what is the equilibrium probability that any individual chosen at random is homozygous for one or other allele?' An elegant answer to this problem has been given by Kimura and Crow (1964). Suppose that the required probability in generation $t$ is $F_t$. Then in generation $t+1$, two genes chosen at random will be identical if (i) they are both copies of the same gene in the previous generation (probability $(2N)^{-1}(1-u)^2$), or (ii) they are copies of different but identical genes in the previous generation (probability $\{1-(2N)^{-1}\}F_t(1-u)^2$). Thus

$$F_{t+1} = (2N)^{-1}\{1+(2N-1)F_t\}(1-u)^2.$$

At equilibrium, $F_t$ and $F_{t+1}$ will be equal; denoting the common value by $F$, it follows that

$$F = \{(2N)(1-u)^{-2}-2N+1\}^{-1} \approx (1+4Nu)^{-1}. \tag{6.36}$$

A number of interesting problems in population genetics can be discussed by using the quantity $F^{-1}$; for this reason the quantity

$4Nu+1$ is often called the effective number of alleles present in the population at any given time.

We now turn to our original problem and discuss the mean of the actual number of alleles present. For this purpose we proceed as follows. The mean number of new alleles created by mutation in any generation is $2Nu$; at equilibrium this number must balance the mean number of 'old' alleles lost by random sampling and/or mutation. If the mean number of different alleles present at any time is $n$ and if the mean time that the line inititated by any new mutant exists is $T$, then the latter quantity must be $n/T$. It follows that the equation

$$n = 2NuT \qquad (6.37)$$

expresses the required quantity $n$ in terms of $T$, so that once an expression for $T$ can be found an expression for $n$ follows immediately. An expression for $T$ can be found as follows. If $x$ represents the frequency of any arbitrary allele $A_k$, the drift and diffusion coefficients for changes in $x$ are found from (4.37) and (6.35) to be

$$m(x) = -\beta x, \qquad v(x) = x(1-x), \qquad (6.38)$$

where $\beta = 2Nu$. Because $x = 0$ is the only absorbing state, the transient function will be defined by (5.39) and (5.40). Since we necessarily have $p = (2N)^{-1}$, it follows that, in terms of generations,

$$T = 2N \int_0^1 t(x)dx, \qquad (6.39)$$

where

$$t(x) = 2x^{-1}(1-2\beta)^{-1}[(1-x)^{2\beta-1}-1], \, 0 < x \leq (2N)^{-1},$$

$$t(x) = 2x^{-1}(1-x)^{2\beta-1}(1-2\beta)^{-1}[1-\{1-(2N)^{-1}\}^{1-2\beta}], \, (2N)^{-1} \leq x < 1.$$

Clearly

$$t(x) \approx 2, \, 0 < x \leq (2N)^{-1}, \qquad (6.40)$$

$$t(x) \approx 2x^{-1}(1-x)^{2\beta-1}(2N)^{-1}, \, (2N)^{-1} \leq x < 1, \qquad (6.41)$$

so that

$$T \approx 2\left[1 + \int_{(2N)^{-1}}^1 x^{-1}(1-x)^{4Nu-1}dx\right].$$

To the same degree of approximation, then,

$$n = 4Nu\left[1 + \int_{(2N)^{-1}}^{1} x^{-1}(1-x)^{4Nu-1}dx\right]. \qquad (6.42)$$

Equation (6.42) expresses the required mean number $n$ in terms of $N$ and $u$; a similar result has been given for a related model by Karlin and McGregor (1966).

It is clearly possible to go further than eqn. (6.42) and to state that if $n(x_1, x_2)$ is the mean number of alleles present at any time whose frequency is between $x_1$ and $x_2$, then

$$n(x_1, x_2) = 4Nu\int_{x_1}^{x_2} x^{-1}(1-x)^{4Nu-1}dx, \quad (2N)^{-1} \leqq x_1 < x_2 \leqq 1. \qquad (6.43)$$

The form of the integrand in eqn. (6.43) makes it possible for us to discuss briefly the frequency distribution of the various alleles. Since the integrand becomes very large at $x = (2N)^{-1}$, there will usually be a comparatively large number of alleles having very low frequency. Among these will no doubt be most of those alleles only recently formed by mutation. If the mutation rate is sufficiently low $(u < (4N)^{-1})$ the integrand becomes infinite at $x = 1$, suggesting that in such cases there will often be one allele having very large frequency, together with a number of other alleles having very low frequency. A larger mutation rate $(u > (4N)^{-1})$ is apparently sufficient to ensure that no allele is able to attain a high frequency. In all cases, alleles having moderate frequency are rare. All these conclusions, except perhaps for the last, agree with what one expects on common-sense grounds.

It is possible to use (6.43) to reobtain (6.36). The probability that two genes drawn at random from the population are both from a given allele whose frequency is $x$ is simple $x^2$. The mean number of alleles whose frequency is between $x$ and $x + \delta x$ is, from (6.43),

$$4Nux^{-1}(1-x)^{4Nu-1}\delta x,$$

so that the probability that the two genes drawn at random are identical is thus

$$4Nu\int_{0}^{1} x^2[x^{-1}(1-x)^{4Nu-1}]dx \approx (1+4Nu)^{-1},$$

in agreement with (6.36).

69

F

While the method outlined above gives us quite quickly an expression for the mean number of different alleles present at any time, it does not seem to generalize to provide further characteristics of the distribution of this number; in particular it does not seem to generalize to provide an expression for the variance of the number of alleles maintained, and indeed it seems very difficult to find a formula for the variance by any sort of mathematical procedure.

To get some idea of the complete distribution, as well as to check the adequacy of (6.42), it is probably necessary to resort to high-speed simulation. As an example of the sort of result to be expected, Table 6.3 records the result of such a simulation for the case $N = 500$, $u = 0.001$, where the population was observed for 800 consecutive 'equilibrium' generations. The data are taken from Ewens and Ewens (1966).

TABLE 6.3. Number of generations for which the number of different alleles assumed the indicated values in 800 consecutive equilibrium generations.

| Number of alleles | Number of generations | Number of alleles | Number of generations |
|:---:|:---:|:---:|:---:|
| 7 | 4 | 15 | 80 |
| 8 | 15 | 16 | 58 |
| 9 | 38 | 17 | 21 |
| 10 | 82 | 18 | 11 |
| 11 | 118 | 19 | 2 |
| 12 | 139 | 20 | 1 |
| 13 | 121 | | |
| 14 | 110 | Total | 800 |

The mean number of different alleles maintained in any generation, for the data of Table 6.3, is 12·66. This is within reasonable sampling fluctuations of the value predicted by (6.42), namely 13·9. It appears further that the number of different alleles in any generation should not differ radically from the mean; Table 6.3 suggests that for the numerical example considered, it is unlikely that at any given time this number will fall outside the range 9–16.

The numerical simulation can be extended quite easily to provide an assessment, parallel to that of Section 4.5, of the effect of geographical subdivision. Table 6.4 below records the result of a simulation identical to that discussed above, except that the population was divided into 25 sub-populations of 20 individuals each. With probability 0·904 an individual mated with an individual from the same population and with probability 0·096 mated with an individual from some other subpopulation. It is found for the results given in Table 6.4 that the mean number of different alleles in any generation is 13·52, which differs but little from the number observed and predicted for the random-breeding case. This confirms our previous conclusion that geographical inbreeding has a very small effect on population behaviour.

TABLE 6.4. Number of generations for which the number of alleles assumed the indicated values in 800 consecutive equilibrium generations (inbred case).

| Number of alleles | Number of generations | Number of alleles | Number of generations |
|---|---|---|---|
| 5 | 2 | 15 | 86 |
| 6 | 1 | 16 | 73 |
| 7 | 13 | 17 | 69 |
| 8 | 30 | 18 | 47 |
| 9 | 45 | 19 | 25 |
| 10 | 55 | 20 | 7 |
| 11 | 71 | 21 | 2 |
| 12 | 70 | | |
| 13 | 94 | | |
| 14 | 110 | Total | 800 |

## 6.5. Self-sterility alleles

The discussion in the previous section is intended mainly to show the lines along which we should attempt to solve a considerably more complex question, namely that of the problem of self-sterility alleles. In a number of plant populations it has been observed that an $A_iA_j$ plant cannot be pollinated by pollen carrying either an $A_i$ or an $A_j$ gene. One effect of this is that homozygotes cannot occur; a consequence is that the Hardy-Weinberg proportions cannot hold. A second and more important effect is that rare genes have a 'quasi-selective advantage', for they are inhibited on a smaller number of

71

plants than more commonly occurring genes, which correspondingly suffer a 'quasi-selective disadvantage'. A further trivial consequence is that no gene can appear with frequency more than $\frac{1}{2}$.

It has been observed that self-sterility populations, even when small, often contain an exceedingly large number of different alleles. As an example the species *Oenothera organensis* is believed to consist entirely of one small population of size about 500 restricted to a small mountain area in New Mexico. In this population, 45 different alleles have been observed in a single generation. Clearly the large number is explained in part by the quasi-selective advantage of rare alleles; however one may still ask what mutation rate, assuming all mutations form entirely new alleles, is necessary to explain the presence of such a large number of alleles. Equation (6.37) shows that this question can be answered once an expression for $T$, the mean number of generations for which the line initiated by any new mutant survives, can be found. Further, if diffusion methods are to be used to approximate $T$, all that is required is an expression for both $m(x)$ and $v(x)$, where these are respectively the mean change and the variance of the change of the frequency of any allele whose current frequency is $x$.

Unfortunately, it is impossible to provide expressions for $m(x)$ and $v(x)$ in terms of $x$ alone. This is because the number of genes of any given allelic type is not a Markovian variate (unlike the corresponding number in the example considered in the previous section). If there are $k$ alleles present at any one time, a complete analysis must be carried out in terms of the joint behaviour of the $\frac{1}{2}k(k-1)$ possible genotypes. With $k = 45$ this would mean joint consideration of 990 variables. Clearly this is impossible and from the start we must be content with approximative methods.

Before considering such approximative methods it is useful to write down explicitly the model we shall use. We consider a population admitting a fixed number $N$ of individuals in each generation. An individual is chosen at random and one of the two sterility genes is taken at random. In the same way, a second gene is then chosen from some other individual. If this second gene is identical to one or other of the two genes from the first individual, the mating is classed as a failure and both individuals are replaced in the population. If not, the mating is a success and the two genes so chosen form an individual for the daughter generation. After the formation of this individual, each sterility gene can mutate, with probability $u$, to an entirely new sterility gene. Both parents are then replaced in the population and the whole procedure repeated $N$ times to make up the daughter generation. We shall call this procedure, which we choose because it is readily adapted for computer simulation, and despite its inadequacies as a reflection of the real world, the 'self-sterility model'.

Suppose that at any one time there are $k$ alleles, $A_1, \ldots, A_k$ present in the population. We consider the frequency $x$ of any one of these, say $A_1$, and to make progress make the following assumptions: (i) that the frequency of $A_i A_j$ is the product of the frequencies of $A_i$ and $A_j$, (ii) that $A_2, \ldots, A_k$ have equal frequencies, namely $(1-x)/(k-1)$.

Consider now the frequency $x'$ of $A_1$ in the following generation. Ignoring for the moment changes in $x$ due to mutation, $x'$ is made up from contributions from the ovules and the pollen of the previous generation, the contribution from the former being $\frac{1}{2}x$. The contribution from pollen is a little more complicated. $A_1$ pollen can function only on non-$A_1$ styles, the frequency of which is $1-2x$. For such styles, $A_1$ pollen has a more than proportional chance of functioning, because for each style two other alleles will be inhibited. It follows that the probability that $A_1$ will pollinate such a style is

$$\frac{x}{1-2(k-1)^{-1}(1-x)}.$$

Hence

$$E(x') = \tfrac{1}{2}x + \frac{\tfrac{1}{2}x(1-2x)}{1-2(k-1)^{-1}(1-x)},$$

so that the expected increase $E(x'-x)$ in the frequency of $A_1$ is

$$E(x'-x) = \frac{x(1-kx)}{k-3+2x}. \tag{6.44}$$

Clearly this is positive for $x < k^{-1}$, demonstrating the amount of the selective advantage of rare alleles. To a similar approximation the variance of $x'-x$ turns out to be

$$\text{Var}(x'-x) = (2N)^{-1}x(1-x). \tag{6.45}$$

Note that while this happens to be identical to the variance formula obtained from eqn. (4.1), such identity is coincidental and eqn. (6.45) in no way implies a binomial distribution for $2Nx'$; indeed the model under consideration is far too complicated for such a simple distribution to apply. A consequence of this is that eqn. (6.45) in no way contradicts the fact that $x$ cannot exceed $\frac{1}{2}$. Using the theory of Chapter 5, and supposing for the moment that diffusion approximations to the discrete model are sufficiently accurate, the drift and diffusion coefficients for the self-sterility model are

$$m(x) = \frac{2Nx(1-kx)}{k-3+2x} - 2Nux \qquad (6.46)$$

(the latter term arising from the steady mutation to new alleles),

$$v(x) = x(1-x). \qquad (6.47)$$

To a sufficient approximation, (6.46) may be written

$$m(x) = \{2N/(k-3)\}x(1-kx) - 2Nux. \qquad (6.48)$$

Joint use of eqns. (5.40), (6.37), (6.47), and (6.48) now shows that the equilibrium number $k$ of different self-sterility alleles maintained in a population of size $N$ by a mutation rate $u$ to new alleles is given, to the degree of approximation used, by the formula

$$k = 4Nu \int_{(2N)^{-1}}^{1} x^{-1}(1-x)^{4Nu+4N(k-1)/(k-3)-1} \exp\left\{-4Nxk/(k-3)\right\}dx.$$

Since $4N$ is large compared to $4Nu$ and to 1, a negligible loss of accuracy is incurred by replacing this by the simpler formula

$$k = 4Nu \int_{2N^{-1}}^{1} x^{-1}(1-x)^{4N(k-1)/(k-3)} \exp\left\{4Nxk/(k-3)\right\}dx. \qquad (6.49)$$

An even simpler formula than this can be found. The above derivation shows that, at equilibrium, the mean number of alleles having frequency in the range $(x, x+\delta x)$ is $n(x)\delta x$, where

$$n(x) = 4Nux^{-1}(1-x)^{4N(k-1)/(k-3)} \exp\left\{4Nxk/(k-3)\right\}.$$

Since

$$\int_{0}^{1} xn(x)dx = 1,$$

eqn. (6.49) can be reduced to

$$1 = 4Nu \int_{0}^{1} (1-x)^{4N(k-1)/(k-3)} \exp\left\{4Nxk/(k-3)\right\}dx, \qquad (6.50)$$

which is easier to use for numerical computation than is (6.49).

In the case $N = 500$, $u = 0·001$, the solution of eqn. (6.50) (which must be found numerically by trial and error) is approximately $k = 30$. Before accepting this value, however, it must be remembered that it was derived after making a considerable number of approximations. Other than those explicitly mentioned, it should be pointed out that with $N = 500$, $k = 30$, $u = 0·001$, the expression (6.48) for $m(x)$ is

$$37\, x\, (1 - 30x) - x$$

Diffusion methods apply only when this is of the same order of magnitude as $v(x)$, and it is not at all obvious that this will be the case.

It is therefore necessary to check the adequacy of eqn. (6.50) by a Monte Carlo simulation. This has been done by several authors, and as a first example some empirical values corresponding to those of Tables 6.3 and 6.4 are presented below in Table 6.5. Values for the geographical inbreeding case, the details of which are identical to those of Table 6.4, are given also.

In the random-mating case, the mean number of different alleles in each generation is $32·18$; for the geographical inbreeding case it is

TABLE 6.5. Number of generations for which the number of alleles assumed the indicated values in 800 consecutive equilibrium generations in a self-sterility model simulation.

| Number of alleles | Number of generations | | Number of alleles | Number of generations | |
|---|---|---|---|---|---|
| | random | inbred | | random | inbred |
| 22 | 2 | 0 | 34 | 93 | 125 |
| 23 | 1 | 0 | 35 | 68 | 91 |
| 24 | 6 | 0 | 36 | 45 | 69 |
| 25 | 5 | 13 | 37 | 31 | 59 |
| 26 | 15 | 26 | 38 | 11 | 23 |
| 27 | 35 | 20 | 39 | 6 | 12 |
| 28 | 62 | 29 | 40 | 8 | 2 |
| 29 | 51 | 37 | 41 | 6 | 0 |
| 30 | 67 | 53 | 42 | 2 | 0 |
| 31 | 65 | 61 | 43 | 2 | 0 |
| 32 | 90 | 65 | | | |
| 33 | 129 | 115 | Total | 800 | 800 |

32·88. Clearly geographical inbreeding is again of negligible importance. The main point, however, is that the mean number of alleles maintained for the random mating case agrees very well with the value predicted by eqn. (6.50).

On the other hand, a more extensive set of calculations (Mayo, (1966)) reveals that such good agreement is not necessarily general. For $N = 250$, $k = 37$, eqn. (6.50) yields $u \approx 0·01$, while Mayo's sampling experiments yield $u \approx 0·004$. For $N = 1000$, $k = 37$, eqn. (6.50) gives $u \approx 0·0006$; here Mayo finds empirically $u \approx 0·0002$. Mayo surmises that the reason for the partial breakdown of eqn. (6.50) is that use of diffusion methods is not quite justified. Further Monte Carlo experiments must be carried out before the range of validity of eqn. (6.50) can be found.

So far as the *Oenothera* population is concerned, a mutation rate of about 0·0014 is required to explain the observed number of alleles. This is, of course, an extremely high mutation rate and does not appear to agree with the extremely low mutation rates found experimentally for this species (Lewis, (1948)). While several hypotheses may now be put forward, none appears to be completely satisfactory and more work in this problem is undoubtedly required.

### 6.6. Comparison with formulae of Fisher

The distributions derived in previous sections have, of course, been known for a long time; they were initially derived in the late 1920's independently by R. A. Fisher and S. Wright. It is therefore of interest to discuss briefly these original analyses, and for this purpose it is sufficient to consider the argument of Fisher (1958, Chapter 4). Consideration of Fisher's analysis provides in addition an indication of how likely diffusion formulae are to differ from the values provided by a discrete analysis.

We consider first the asymptotic conditional distribution of the process governed by (4.1); this distribution was very prominent in the early literature. From eqn. (5.20) we expect that all values of the random variable will have, to a close approximation, the same probability $1/(2N-1)$. Fisher (p. 94) uses an elegant functional equation technique to show that such a conclusion is not quite true for extreme values of the random variable; denoting the asymptotic conditional distribution by

$$(2N-1)^{-1}\{q_1, q_2, \ldots \ldots, q_{2N-1}\}, \tag{6.51}$$

Fisher shows that for small integer $i$,

$$q_i = q_{2N-i} \approx 1 - (6i)^{-1}, \qquad (6.52)$$

rather than $q_i = 1$, as is suggested by (5.20). Indeed for extremely small integers $i$, Fisher provides an even more accurate expression than (6.52). A second formula, which is less accurate than the latter for very small $i$, but more accurate for moderate values of $i$, is (Ewens (1965))

$$q_i = 1 - 2N\{6i(2N-i)\}^{-1} + (2N)^{-1}\{\tfrac{2}{3} + \tfrac{1}{3}\gamma - \tfrac{1}{3}\ln(2N) + \tfrac{1}{3}\ln i(2N-i)\},$$

where $\gamma$ is Euler's constant $0 \cdot 5772 \ldots$

These correction terms are clearly of little genetical significance. Indeed, it is possible to go further and ask whether the asymptotic conditional distribution itself can have much significance. The reason for this is that the spectral expansion (5.19) suggests that the time required until this distribution becomes relevant is so long that absorption at $x = 0$ or $x = 1$ will usually occur beforehand. Thus, for example, if $p = \tfrac{1}{3}$,

$$f(x;t) = \tfrac{4}{3}\exp(-t) + 10(1-2x)\exp(-3t) + \ldots$$

For $x = \tfrac{1}{3}$, say, the first term will dominate the remaining terms only for $t$ larger than about 2, that is to say after about $4N$ generations. Not only is this such a long time that we may be unable to assume constancy of environment; eqn. (6.7) shows that, for $p = \tfrac{1}{3}$, the mean time until $x = 0$ or $x = 1$ is reached is about $2 \cdot 5N$ generations, so that such absorption will usually occur well before the asymptotic conditional distribution becomes relevant.

The second function considered by Fisher (pp. 96–99) is the transient function of (4.1) when the initial value of the random variable is 1. The diffusion approximation to this function is

$$t(x) = 2\{1 - (2N)^{-1}\}/(1-x), \qquad 0 < x \le (2N)^{-1}, \qquad (6.53)$$

$$t(x) = 2\{(2N)^{-1}\}/x, \qquad (2N)^{-1} \le x < 1, \qquad (6.54)$$

yielding a mean absorption time (in terms of generations) of approximately

$$2 + 2\ln 2N. \qquad (6.55)$$

Further, (6.54) shows that, to the same order of approximation, the mean number of generations for which a random variable obeying

(4.1) assumes any value $i$, given that its initial value is 1, is

$$2/i, \qquad i = 1, 2, \ldots, 2N - 1. \tag{6.56}$$

By a second use of functional equations, Fisher is able to obtain slightly more accurate values for small $i$ than are given by (6.56), and thus a slightly more accurate value than (6.55), namely

$$1 \cdot 355 + 2 \ln 2N. \tag{6.57}$$

There is one problem which was of some concern to Fisher and other early writers, namely the non-convergence of (6.54) at $x = 0$. Indeed this was the reason why Fisher resorted to use of functional equations to discuss the behaviour of $t(x)$ near $x = 0$, since it was assumed that, because of the non-convergence of (6.54), diffusion methods did not apply near $x = 0$. This is, in fact, not true, and the fact that a different formula (viz. (6.53)) applies for very small $x$ removes all problems of non-convergence.

# CHAPTER 7

# Results Derived from Branching Processes

## 7.1. Survival probabilities

Some of the results concerning the stochastic behaviour of new mutants obtained in the previous chapter may be re-derived and extended by using branching process theory, and in this chapter we consider how this can be done.

Firstly, since we are dealing with a finite population, it is clear that use of branching processes can give only approximate answers. This is not as serious as might be thought, since we have shown, for example, that survival probabilities of non-recessive new mutants are effectively independent of population size, so long as the latter is large.

Taking this particular case first, suppose that a non-recessive $A_1$ mutant is introduced into a previously pure $A_2A_2$ population. Then homozygotes $A_1A_1$ will not usually appear until the frequency of $A_1$ is quite high; we cannot normally expect the appearance of such a homozygote until the number of $A_1$ genes is about $\sqrt{(2N)}$, where $N$ is the population size. By this time the fate of the new mutant is usually effectively determined. We therefore ignore the possibility of formation of such homozygotes and concentrate attention on the number of heterozygotes $A_1A_2$.

Before going further, it is useful to discuss what is meant by the survival of a favourable new mutant. If the fitness of $A_1A_1$ exceeds that of $A_1A_2$, once the mutant is established in some numbers its frequency will almost certainly increase to unity. On the other hand, if the fitness of $A_1A_2$ exceeds that of $A_1A_1$ (as well as that of $A_2A_2$), then the frequency of $A_1$ will approach a quasi-stable equilibrium between 0 and 1. Since both cases are of interest, we shall take the expression 'survival of a favourable new mutant' to imply the increase in frequency of the mutant to such an extent that the probability of loss of the mutant by random sampling in anything other than an extremely long time is completely negligible. While this definition is not very precise, we shall never encounter any difficulty in using it.

If the population size before the mutation occurred was stable, the fitness of $A_2A_2$ can be taken as unity. We shall suppose that the fitness of $A_1A_2$ is then $1+s$, and for convenience shall refer to such a heterozygote as a mutant. Clearly the mean number of mutant offspring from any mutant is $1+s$; if the probability that a mutant leaves exactly $i$ mutant offspring is $f_i$, we must have

$$f_i \geqq 0, \qquad \sum f_i = 1, \qquad \sum i f_i = 1+s.$$

Writing $f(\theta) = \sum \theta^i f_i$, a well known result from branching process theory states that the probability $\theta$ of extinction of the line initiated by a single mutant is the smallest positive root of the equation

$$\theta = f(\theta). \tag{7.1}$$

This solution is unity when $s \leqq 0$ and is less than unity if $s > 0$. In the particular case of Poisson offspring distribution, this equation becomes

$$\theta = \exp \{(1+s)(\theta-1)\}, \tag{7.2}$$

and writing $\pi = 1 - \theta$, eqn. (7.2) reduces, for small $s$, to

$$\pi \approx 2s. \tag{7.3}$$

For larger $s$ a more accurate expression is

$$\pi \approx 2s - {}_8\, s^2. \tag{7.4}$$

Note that (7.3) agrees with the result (6.3) found by diffusion methods. Since the latter applies for $s$ of the order $N^{-1}$, while branching process methods apply for large $s$, eqn. (7.3) appears to be true for a wide range of values of $s$.

A result more general than (7.3) may be found by expanding, in eqn. (7.1), $f(\theta)$ about the point $\theta = 1$ and ignoring terms of order $(1-\theta)^3$. This gives

$$\pi \approx 2s/\sigma^2, \tag{7.5}$$

where $\sigma^2$ is the variance of the offspring distribution $\{f_i\}$; in this form the dependence of survival probabilities on variance as well as on selective advantage is clearly demonstrated.

It is to be expected that if the mutant is lost as a result of random sampling, such loss will usually occur fairly soon after the mutation first arises, for if the mutant survives for a number of generations it will normally increase in numbers and thus be comparatively safe from random extinction. As an example of this, Fisher (1930) has computed a table indicating, in the case $s = 0\cdot01$, the probability that the line initiated by any mutant still survives after a given number of generations. Thus the probability of eventual extinction of such a mutant is $0\cdot9803$; the probability that such extinction has occurred by the 15th generation is $0\cdot8783$. It would be possible to compute, by diffusion methods, the mean time for extinction given that eventual extinction occurs, but the calculations become rather involved and will be omitted here.

## 7.2. Multiple alleles

A more complicated situation arises when a new mutation occurs in a population in which exist alleles $A_1, A_2, \ldots A_k$, each with non-negligible frequency. Suppose that the fitness of $A_i A_j$ individuals is $w_{ij}$, the frequency of $A_i$ is $p_i$, and that genotype frequencies are in Hardy-Weinberg form. Thus the mean fitness of the population before the mutation is $W = \sum w_{ij} p_i p_j$, which, in a stable population, may be taken as unity. With this convention, if the new mutant is denoted by $A_{k+1}$, we denote the fitness of $A_i A_{k+1}$ by $\mu_i$.

In deriving the survival probability of $A_{k+1}$, we shall suppose that the possibility of formation of homozygous mutants can be ignored, and will then describe an $A_i A_{k+1}$ individual as a mutant of type $i$. If in generation $t$ the number of type $i$ mutants is $n_i(t)$, then clearly

$$En_j(t+1) = p_j \sum_i \mu_i n_i(t), \qquad j = 1, \ldots, k. \qquad (7.6)$$

In matrix terms this may be written

$$En(t+1) = Mn(t), \qquad (7.7)$$

where

$$M = p\mu', p' = (p_1, \ldots, p_k), \qquad \mu' = (\mu_1, \ldots, \mu_k). \qquad (7.8)$$

To obtain survival probabilities, it is necessary to know not only the mean number of mutants (given by (7.6)), but the complete joint distribution of the number of $-A_{k+1}$ offspring from any $A_i A_{k+1}$ parent. If the joint generating function of this distribution is

$f_i(\theta_1, \ldots \ldots, \theta_k)$, then $f_i(1, \ldots, 1) = 1$ and writing $f_i(\theta, \ldots, \theta)$
$= f_i(\theta)$,

$$f_i'(1) = \mu_i.$$

Now any mutant offspring from an $A_i A_{k+1}$ parent will be $A_j A_{k+1}$ with probability $p_j$; from this it follows that $f_i(\theta_1, \ldots, \theta_k)$ assumes the special form

$$f_i(\theta_1, \ldots, \theta_k) = f_i(p_1\theta_1 + \ldots + p_k\theta_k). \qquad (7.9)$$

The following theorem, taken from Harris (1963, p. 41), can now be applied.

*Theorem* 7.1. For a multiple type branching process governed by (7.7) (that is, for a set of objects of various types, each of which can produce offspring of any type according to branching process laws), the probability of survival of the mutant $A_{k+1}$ is positive if and only if the largest eigenvalue of $M$ exceeds unity. In this case, if $\theta_i$ is the probability of extinction of the mutant given a single initial mutant of type $i$, the $\theta_i$ are the unique positive solutions (less than unity) of the set of equations

$$\theta_i = f_i(\theta_1, \ldots \ldots, \theta_k), \qquad i = 1, \ldots, k. \qquad (7.10)$$

In our case, eqns. (7.9) and (7.10) yield jointly

$$\theta_i = f_i(p_1\theta_1 + \ldots \ldots + p_k\theta_k), \qquad i = 1, \ldots, k. \qquad (7.11)$$

Before considering the solution of eqn. (7.11), it is useful to consider first the condition that the various $\theta_i$ be less than unity. The form (7.8) of $M$ shows that $M$ is of rank unity, and thus has only one non-zero eigenvalue. This eigenvalue must then be identical to the sum of all eigenvalues, which is trace $(M)$ or $\sum p_i\mu_i$. Thus the necessary and sufficient condition that survival probabilities be positive is

$$\sum p_i\mu_i > 1. \qquad (7.12)$$

When (7.12) holds, the easiest way to solve the system (7.11) of equations is to multiply the $i^{th}$ equation by $p_i$ and add over $i$. Defining $\theta = \sum p_i\theta_i$, this leads to

$$\theta = \sum p_i f_i(\theta). \qquad (7.13)$$

Often $\theta$ is really the quantity of interest, since if we do not know the exact genotype formed by the initial mutant, it is reasonable to assume that this genotype is $A_i A_{k+1}$ with probability $p_i$. In any event, individual $\theta_i$ values are best found by solving (7.13) for $\theta$ and then calculating each $\theta_i$ directly from eqn. (7.11).

While it is difficult to draw general conclusions from this result, it is possible to find an analogue of (7·5) when $\mu = \sum p_i \mu_i$ is only slightly greater than unity. If $\sigma_i{}^2$ is the variance of the number of mutant offspring from type $i$ mutants, and if $\sigma^2 = \sum p_i \sigma_i{}^2$, eqn. (7.13) yields, on expanding the right-hand side about $\theta = 1$ and ignoring terms of order $(1-\theta)^3$,

$$\pi = 1 - \theta \approx 2(\mu - 1)/\sigma^2. \qquad (7.14)$$

Application of eqn. (7.11) then gives

$$\pi_i = 1 - \theta_i \approx \pi \mu_i - \tfrac{1}{2}\pi^2 \sigma_i{}^2, \qquad i = 1, \ldots, k. \qquad (7.15)$$

If it is assumed further that all offspring distributions are Poisson, further results can be found. Equation (7.13) can be rewritten

$$1 - \pi = \sum p_i \exp\left(-\mu_i \pi\right), \qquad (7.16)$$

while the set of equations (7.11) becomes

$$1 - \pi_i = \exp\left(-\mu_i \pi\right). \qquad (7.17)$$

At least two conclusions can be drawn from this result. Firstly if, as a rough approximation, the multivariate nature of the process had been ignored and a univariate analysis carried out under the supposition that the fitness of the mutant was $\sum p_i \mu_i$, then the survival probabilities $\pi^*$ so calculated would satisfy the equation

$$1 - \pi^* = \exp\left(-\pi^* \sum p_i \mu_i\right). \qquad (7.18)$$

This is again positive only when (7.12) holds, but it is easy to show that in this case $\pi^*$ necessarily exceeds the true weighted survival probability $\pi$ (except in the limiting case $\mu_1 = \mu_2 = \ldots \ldots = \mu_k$, when they coincide). A somewhat over-optimistic value for survival probability is thus obtained under this method.

Secondly, eqn. (7.16) shows that $\pi$ is not a linear function of $\sum p_i \mu_i$. Indeed $\pi$ need not even be an increasing function of $\sum p_i \mu_i$. This means that it is possible to produce two cases, for the first of which the value of $\sum p_i \mu_i$ exceeds that for the second, but for which the value of $\pi$ is larger in the second case. We provide the following example:

Case A:

$$\mu_1 = \mu_2 = 1 \cdot 5, \qquad p_1, p_2 \text{ arbitrary.}$$

Here

$$\sum p_i \mu_i = 1 \cdot 5, \qquad \pi = 0 \cdot 58281.$$

Case B:

$$\mu_1 = 1, \quad \mu_2 = 2, \quad p_1 = 0 \cdot 429, \quad p = 0 \cdot 571.$$

Here

$$\sum p_i \mu_i = 1 \cdot 571,$$

but

$$\pi_1 = 0 \cdot 44122, \quad \pi_2 = 0 \cdot 68776, \quad \pi = 0 \cdot 5820.$$

It is clear that use of the approximate method can sometimes give a misleading comparison between survival probabilities in any two situations; on the other hand one feels that in the great majority of cases, increasing $\sum p_i \mu_i$ increases $\pi$.

### 7.3. Fluctuating environments

We return in this section to the case of single $A_1$ mutant gene in a previously pure $A_2 A_2$ population; as before, the possibility of formation of $A_1 A_1$ individuals can be ignored and an $A_1 A_2$ individual can be described as a mutant.

The results of the previous sections assume constancy of population size, and the purpose of this section is to generalize them to two important cases where this assumption is no longer made. These generalizations are (i) when the population size assumes a cyclic sequence of values $N_1, N_2, \ldots, N_k, N_1, N_2, \ldots$ and (ii) when the size of the population increases from a comparatively low value to a rather higher value, where it levels off.

Taking firstly the cyclic population size, it is clear that the probability of survival of a mutant must depend to some extent on the size of the population when the initial mutant is born. Thus any mutant (i.e. not only the initial mutant, but also all its mutant descendants) must be classified into one or other of $k$ types so that a mutant is of type $i$ if it is born when the population size is $N_i$ ($i = 1, \ldots, k$). Clearly a type $i$ offspring can have only type $i+1$ offspring, while if $A_1 A_2$ has fitness $1+s$ and $A_2 A_2$ has unit fitness, the mean number of such offspring is $(1+s) N_{i+1}/N_i$. (We make the notational convention type $(k+1) = \text{type } 1$, $N_{k+1} = N_1$.) With this convention, the theory of multiple type branching processes outlined in the previous section can be applied. If $n_i(t)$ refers to the number of type $i$ mutants in generation $t$, then eqn. (7.7) holds if we put

$$M = (1+s)\begin{pmatrix} 0 & N_2/N_1 & 0 & 0\dots\dots0 \\ 0 & 0 & N_3/N_2 & 0\dots\dots0 \\ 0 & 0 & 0 & N_4/N_3\dots\dots0 \\ \vdots & \vdots & \vdots & \vdots & \vdots \\ N_1/N_k & 0 & 0 & 0\dots\dots0 \end{pmatrix}$$

The eigenvalues of $M$ are $(1+s)\psi_j(j = 1, \dots, k)$, where the $\psi_j$ are the $k^{th}$ roots of unity; the condition that survival probabilities be positive is thus $s > 0$. When this is the case, eqn. (7.10) can be written

$$\theta_i = f_i(\theta_{i+1}), \qquad i = 1, \dots, k, \tag{7.19}$$

where $f_i(\theta)$ is the generating function of the number of mutant off-spring from a mutant parent living when the population size is $N_i$.

To obtain more explicit results, it will be assumed that all such offspring distributions are Poisson. Considering survival probabilities $\pi_i = 1 - \theta_i$, eqn. (7.19) reduces to

$$-ln(1 - \pi_i) = (1+s)(N_{i+1}/N_i)\pi_{i+1}, \quad i = 1, \dots k, \tag{7.20}$$

with $\pi_{k+1} = \pi_1$. Equation (7.20) must be solved numerically, and a useful algorithm for this is as follows. Choose a trial value $\pi_1$ and from eqn. (7.20) with $i = 1$, compute $\pi_2$. Using this value, use eqn. (7.20) with $i = 2$ to compute $\pi_3$, and so on. If the value $\pi_{k+1}$ eventually obtained exceeds the trial value $\pi_1$, the latter is too large and a smaller trial value should be chosen; if $\pi_{k+1}$ is less than $\pi_1$ then $\pi_1$ is too small. If $\pi_{k+1} = \pi_1$, then this is the required value and the remaining $\pi_i$ will have been calculated.

As an example, if the population size assumes the sequence of values $N, 2N, 4N, 2N, N, 2N, \dots\dots$, and if $s = 0 \cdot 1$, the system (7.20) of equations yields

$$\pi_1 = 0 \cdot 2985, \quad \pi_2 = 0 \cdot 1612, \quad \pi_3 = 0 \cdot 0799, \quad \pi_4 = 0 \cdot 1514.$$

Clearly there is quite a strong dependence of survival probability on population size; indeed survival probabilities seem to be roughly inversely proportional to the size of the population when the mutation first appears. This can be shown, for small $s$, to be a general result, the method of proof being to expand each equation in (7.17) and to

G

take leading terms only (in the same way that eqn. (7.3) was obtained from (7.1)). This method gives

$$\pi_i \approx 2sN^*/N_i, \tag{7.21}$$

where $N^*$ is the harmonic mean of the $N_i$. Equation (7.21) not only confirms our previous observation, but suggests also a further development. It is reasonable to suppose that any mutation which occurs during any cycle will occur when the population size is $N_i$ with probability $N_i/(N_1 + \ldots + N_k)$. In this case the weighted survival probability $(\sum \pi_i N_i)/(\sum N_i)$ can be written, using (7.21), as

$$\pi \approx 2sN^*/\bar{N},$$

where $\bar{N}$ is the arithmetic mean of the $N_i$. For given $k$ and $\bar{N}$, $\pi$ is maximized when the $N_i$ are equal; if the $N_i$ differ considerably, the value of $\pi$ will be rather less than this maximum value.

The second situation to be considered is that where the population size assumes the sequence of values $N_1, N_2, \ldots, N_{k-1}, N_k, N_k, Nk$, $\ldots\ldots$ Here we are thinking mainly of the case of a small population which finds a favourable environment and grows in size until saturation is reached. Thus we have in mind mainly the case $N_i \geqq N_{i-1}$, although the subsequent discussion is quite general.

Again a mutant is classified into one or other of $k$ different types, depending on the size of the population when it is born. If it is supposed that, on the average, the mutant $A_1 A_2$ produces $1+s$ mutant offspring for every offspring produced by $A_2 A_2$, the matrix $M$ in eqn. (7.7) becomes

$$M = (1+s) \begin{pmatrix} 0 & N_2/N_1 & 0 & 0\ldots\ldots & 0 \\ 0 & 0 & N_3/N_2 & 0\ldots\ldots & 0 \\ \vdots & & & & \\ 0 & 0 & 0 & 0\ldots\ldots N_k/N_{k-1} \\ 0 & 0 & 0 & 0\ldots\ldots & 1 \end{pmatrix}$$

Again the condition that survival probabilities be positive is $s > 0$; in this case the probabilities $\pi_i$ of survival of the line initiated by a mutant born when the population size is $N_i$ satisfy the system of equations

$$-ln(1-\pi_i) = (1+s)(N_{i+1}/N_i)\pi_{i+1}, \quad i = 1, \ldots, k-1, \tag{7.22}$$

$$-ln(1-\pi_k) = (1+s)\pi_k. \tag{7.23}$$

Here the $\pi_i$ are found by first solving eqn. (7.23) and then working backwards in eqn. (7.22); in this case no iterative procedure is required.

As a numerical example, suppose that the population size assumes the sequence of values $N, 2N, 4N, 8N, 8N, 8N, \ldots$, and that $s = 0 \cdot 1$. Then, from eqn. (7.23), $\pi_4 = 0 \cdot 1761$. Use of eqn. (7.22) with $i = 3$ yields $\pi_3 = 0 \cdot 3212$, and similarly $\pi_2 = 0 \cdot 5067$, $\pi_1 = 0 \cdot 6720$. Clearly the increase in population size has the effect that the probability of survival of a mutant born in the early generations is quite substantial. In this respect it is interesting to note that had the population size increased immediately from $N$ to $8N$, the probability of survival of a mutant born in the first generation would be $0 \cdot 7877$, which is rather more than the value found when the increase in size requires three generations. Presumably this is typical of a general result.

It is interesting to consider briefly the fate of unfavourable new mutants; this topic is treated at greater length in the following section. In the numerical case just considered, any mutant having fitness between $\frac{1}{2}$ and 1 would tend to increase in numbers during the early generations, but would usually die out quite quickly once the population size stabilizes. Thus a mutant with fitness $0 \cdot 75$ born in the first generation has probability $0 \cdot 6438$ of leaving mutant descendants in the fourth generation; in a stable population the corresponding probability is only $0 \cdot 1564$. It follows that while the population size increases we can expect a variety of new and perhaps unusual types to appear, to disappear again fairly soon after the population size stabilizes.

The results of this section make it clear that population size fluctuations can make substantial changes to the probability of survival of new mutants; for this reason, all classical results for survival probabilities assume stable population sizes. For additional results in this direction, see Kojima and Kelleher (1962).

A similar problem to that considered in this section arises if the population is divided into a number of subpopulations, with different fitnesses for the mutant in each. Here what is important is not when the mutation arises, but where it arises. Survival probabilities have been calculated for this situation by Pollack (1966).

## 7.4. Selectively disadvantageous mutants

In this section we consider a population in which the fitness of $A_2 A_2$ individuals exceeds that of $A_1 A_2$ individuals, which in turn exceeds that of $A_1A_1$. Because of mutation $A_2 \rightarrow A_1$, there will exist a small number of $A_1A_2$ individuals in the population; with negligible loss of accuracy it can be assumed that $A_1A_1$ individuals never occur. Clearly the number of $A_1A_2$ individuals is not fixed, and in this section various characteristics of the distribution of this number will be discussed.

Suppose that the fitness of $A_2A_2$ is 1, and that of $A_1A_2$ is $1-s$, where $s$ is assumed positive. (This involves a slight change in notation from previous sections.) Suppose also that the probability that any $A_1A_2$ individual produces exactly $i$ $A_1A_2$ offspring is $p_i$, ($i = 0,1,2, \ldots$), that the population size is $N$ and that $A_2$ mutates to $A_1$ at rate $v$. Back mutation from $A_1$ to $A_2$ can be ignored and, to a sufficiently close approximation, it may be assumed that the number of $A_2$ genes in the population at any time is $2N$. This implies that if the distribution of the number of new mutants derived in any generation from $A_2A_2$ parents is Poisson, (as is most reasonable), then the generating function of the number of such new mutants is $\exp\{2Nv(\theta-1)\}$.

Suppose now that, at equilibrium, the probability that there are exactly $i$ $A_1A_2$ individuals is $h_i(i = 0,1,2, \ldots)$ and write $h(\theta) = \sum h_i\theta_i$. The $A_1A_2$ individuals making up any generation may be divided into two groups, firstly those having an $A_1A_2$ parent, and secondly those new mutants whose parents were $A_2A_2$. At equilibrium, the generating function of distribution of the number in the former category is $h(p(\theta))$ and in the latter $\exp\{2Nv(\theta-1)\}$. Since the two numbers are independent to the degree of approximation used, the distribution of the total number of mutants in this generation is the product of these two generating functions. It follows that $h(\theta)$ satisfies the functional equation

$$h(\theta) = h(p(\theta)) \exp\{2Nv(\theta-1)\}. \qquad (7.24)$$

Even if $p(\theta)$ is known, it is usually difficult to solve this equation for $h(\theta)$. On the other hand, moments of $\{h_i\}$ can be found by differentiating throughout in eqn. (7.24). Thus if the mean and variance of $\{h_i\}$ are $\mu$, $\sigma^2$, then we find

$$\mu = 2Nv/s, \qquad \sigma^2 = 2Nv\{1+\sigma_0^2/s\}/\{2s-s^2\}. \qquad (7.25)$$

where $\sigma_0^2$ is the variance of the distribution $\{p_i\}$.

It is of some interest to compare (7.25) with the analogous diffusion theory results (6.30). The latter were derived under the assumption that $s$ is $0(N^{-1})$, while the former apply best for large $s$. Clearly, when allowance is made for notational changes, the mean values agree. To compare variances it is necessary to put $\sigma_0{}^2 = 1$, since (6.30) is found under this (implicit) assumption. Clearly, for values of $s$ less than about 0·1, the variances are effectively equal. It is therefore reasonable to hope that the complete distribution $\{h_i\}$ is close to (6.29). This result, together with the agreement of (6.3) and (7.3), suggests strongly that, while branching process theory applies best for large $s$, and diffusion theory is strictly valid only when $s$ is of order $N^{-1}$, the range of application of diffusion results is far wider than would initially appear. Supporting evidence for this view is afforded by the results of Section 6.2 and 6.5; nevertheless the problem of the actual range of validity of diffusion formulae is an open one and care must always be exercised in applying diffusion methods.

# CHAPTER 8

# Two-locus Behaviour

## 8.1. Introduction

All the results found in the previous chapters have been derived under the assumption that the fitness of any individual depends solely on his genetic constitution at a single locus. This is, of course, a gross simplification in general, although in many practical problems it is extremely useful. In this and subsequent chapters it is assumed that the fitness of any individual depends on his genetic constitution at two loci. Although this assumption is often hardly less unreal than that made previously, it does allow genuine progress to be made in two directions. Firstly, it enables us to measure the errors committed in treating what is strictly a two-locus problem by a composition of single locus methods; any results found will be relevant to the question of treating what are essentially multi-locus problems by single locus methods. Secondly, two-locus analyses enable us to discuss the evolutionary effect of linkage between loci; we shall find that linkage is a most important factor in population genetics, and indeed all of Chapter 9 is devoted to considering some of its consequences.

Nearly all the analysis which follows is deterministic. This is because a stochastic treatment appears to be very difficult; in particular it appears that there is considerable difficulty in formulating diffusion models which take adequate account of the phenomenon of linkage. On the other hand, previous chapters have given us some idea of when it is possible to ignore stochastic fluctuations, so that we should be able to judge which results we obtain deterministically for two loci will stand unchanged even when stochastic fluctuations are allowed.

## 8.2. Changes in gamete frequency

We consider a population for which the fitness of any individual depends solely on his genetic constitution at two loci, '$A$' and '$B$', at which occur alleles $A_1, A_2$ and $B_1, B_2$ respectively. The recombination fraction between these loci, that is to say the probability that there will be an odd number of crossings-over between them, is assumed to be $R$. It is assumed that $0 \le R \le \frac{1}{2}$; the upper bound $R = \frac{1}{2}$ is achieved if the loci are unlinked (i.e. on different chromosomes).

Clearly 10 different zygotes are possible, namely $A_1B_1/A_1B_1$, $A_1B_1/A_1B_2$, $A_1B_2/A_1B_2$, $A_1B_1/A_2B_1$, $A_1B_1/A_2B_2$, $A_1B_2/A_2B_1$, $A_1B_2/A_2B_2$, $A_2B_1/A_2B_1$, $A_2B_1/A_2B_2$ and $A_2B_2/A_2B_2$, and the frequencies of these in any given generation will be denoted by $y_1, \ldots\ldots,$ $y_{10}$. By considering all possible matings, together with their respective frequencies and outputs, it is found that if no selective differences exist, and if $y_1'$ denotes the frequency of $A_1B_1/A_2B_2$ in the following generation, then

$$y_1' = (y_1 + \tfrac{1}{2}y_2 + \tfrac{1}{2}y_4 + \tfrac{1}{2}y_5(1-R) + \tfrac{1}{2}y_6R)^2, \qquad (8.1)$$

with similar expressions holding for the remaining frequencies. Now the frequency of the gamete $A_1B_1$ forming this following generation is

$$c_1' = y_1 + \tfrac{1}{2}y_2 + \tfrac{1}{2}y_4 + \tfrac{1}{2}y_5(1-R) + \tfrac{1}{2}y_6R, \qquad (8.2)$$

so that eqn. (8.1) may be rewritten

$$y_1' = (c_1')^2. \qquad (8.3)$$

It is not difficult to see that the remaining frequencies $y_i'(i = 2, 3, \ldots, 10)$ are all given by equations analogous to (8.3). Thus, for example, if the frequencies of the gametes $A_2B_1, A_1B_2$ and $A_2B_2$ are denoted $c_2, c_3, c_4$ respectively, then

$$y_4' = 2c_1'c_2'. \qquad (8.4)$$

The set of equations for $y_1', \ldots\ldots, y_{10}$ typified by eqns. (8.3) and (8.4) shows that the following generation of zygotes can be thought of as being made up by 'random sampling of gametes'. This is a most useful and simplifying observation and will be used often; it is the extension of the result derived in Section 1.2. Clearly, under random mating, there is no need to consider the frequencies $y_1, \ldots\ldots, y_{10}$; it is sufficient to consider the four frequencies $c_1, \ldots\ldots, c_4$. (Indeed any three of $c_1, \ldots, c_4$ will be sufficient, but for symmetry we always consider all four.) Using now the frequencies $c_i$, the relations typified by eqns. (8.3) and (8.4) become

$$c_1' = c_1 + R(c_2 c_3 - c_1 c_4), \tag{8.5}$$

$$c_2' = c_2 - R(c_2 c_3 - c_1 c_4), \tag{8.6}$$

$$c_3' = c_3 - R(c_2 c_3 - c_1 c_4), \tag{8.7}$$

$$c_4' = c_4 + R(c_2 c_3 - c_1 c_4). \tag{8.8}$$

Several elementary conclusions may be drawn from these equations. Firstly, it is clear that $c_1' + c_2' = c_1 + c_2$ irrespective of the degree of linkage between the loci. This confirms the previous result (derived under single locus theory) that the frequency of $B_1$ remains unchanged in successive generations; a similar result holds for $A_1$. Secondly, if $c_2 c_3 = c_1 c_4$, then $c_i' = c_i$ for all $i$. The equation

$$c_1 c_4 = c_2 c_3 \tag{8.9}$$

is equivalent to

frequency of $(A_i B_j)$ = (frequency of $A_i$) × (frequency of $B_j$),  (8.10)

and it is important to consider the circumstances under which eqn. (8.10) holds. Since

$$c_1' c_4' - c_2' c_3' = (1 - R)(c_1 c_4 - c_2 c_3), \tag{8.11}$$

it follows that if $R > 0$,

$$c_1(t) c_4(t) - c_2(t) c_3(t) \to 0 \text{ as } t \to \infty. \tag{8.12}$$

Thus eqn. (8.9) should hold to a sufficient approximation after a number of generations; for $R = \frac{1}{2}$, (8.9) will be established very closely after a comparatively small number of generations. Thus, while that part of the Hardy-Weinberg law which states that genotypic frequencies stabilize after one generation of random mating has no analogue for two loci, this is not particularly serious and the main evolutionary consequence of the Mendelian system, namely the intrinsic conservation of variation, still holds.

When eqn. (8.9) holds, the two loci are said to be in 'linkage equilibrium'; otherwise, the quantity $c_1 c_4 - c_2 c_3$ is called the 'coefficient of linkage disequilibrium'. We shall see later that this terminology is a little misleading and will consider an alternative name for the quantity $c_1 c_4 - c_2 c_3$.

## 8.3. The effect of selection

We now consider the case where selective differences between genotypes exist; here the results found in the previous section do not necessarily hold.

While for two loci there are ten possible zygotes, this number is reduced to nine if the assumption is made that the coupling double heterozygote $(A_1B_1/A_2B_2)$ and the repulsion double heterozygote $(A_1B_2/A_2B_1)$ are all in relevant respects identical. We do this for simplification only; the evolutionary consequences of this assumption's not holding are important and have been examined by Parsons (1963).

Under this assumption, suppose that the fitnesses of the nine different zygotes are as displayed in the following matrix:

|          | $A_1A_1$ | $A_1A_2$ | $A_2A_2$ |          |
|----------|----------|----------|----------|----------|
| $B_1B_1$ | $w_{11}$ | $w_{12}$ | $w_{13}$ |          |
| $B_1B_2$ | $w_{21}$ | $w_{22}$ | $w_{23}$ | (8.13)   |
| $B_2B_2$ | $w_{31}$ | $w_{32}$ | $w_{33}$ |          |

It is clear that zygotic frequencies immediately after the formation of the zygotes of any generation can be found by using gametic frequencies, using the argument of random union of gametes (which applies irrespective of selective effects). (Note that this does not hold at later stages in the life cycle, and for this reason we shall compute frequencies at this particular time.) Clearly, at this time, the frequency of $A_1A_1B_1B_1$ is $c_1{}^2$, with similar results for the remaining zygotes, while the mean fitness $W$ of the population is

$$W = w_{11}c_1{}^2 + 2w_{12}c_1c_2 + w_{13}c_2{}^2 + 2w_{21}c_1c_3 + 2w_{22}c_1c_4 + 2w_{22}c_2c_3$$
$$+ 2w_{23}c_2c_4 + w_{31}c_3{}^2 + 2w_{32}c_3c_4 + w_{33}c_4{}^2. \qquad (8.14)$$

Consideration of the various possible matings, together with their frequencies and outputs, shows that the gametic frequencies $c_i{}'$ in the following generation are

$$c_1{}' = W^{-1}[w_{11}c_1{}^2 + w_{12}c_1c_2 + w_{21}c_1c_3 + w_{22}c_1c_4 + Rw_{22}(c_2c_3 - c_1c_4)],$$
$$\qquad (8.15)$$

$$c_2{}' = W^{-1}[w_{12}c_1c_2 + w_{13}c_2{}^2 + w_{22}c_2c_3 + w_{23}c_2c_4 - Rw_{22}(c_2c_3 - c_1c_4)],$$
$$\qquad (8.16)$$

$$c_3{}' = W^{-1}[w_{21}c_1c_3 + w_{22}c_2c_3 + w_{31}c_3{}^2 + w_{32}c_3c_4 - Rw_{22}(c_2c_3 - c_1c_4)],$$
$$\qquad (8.17)$$

$$c_4{}' = W^{-1}[w_{22}c_1c_4 + w_{23}c_2c_4 + w_{32}c_3c_4 + w_{33}c_4{}^2 + Rw_{22}(c_2c_3 - c_1c_4)]$$
$$\qquad (8.18)$$

These formulae generalize eqns. (8.5)–(8.8), to which they reduce when the $w_{ij}$ are equal. They are most important and will form the basis of much of the subsequent development; they are due essentially to Kimura (1956) and explicitly to Lewontin and Kojima (1960).

It is important to examine, firstly, the nature of the equilibrium points associated with eqns. (8.15)-(8.18), and secondly the behaviour of $W$. It is expected from previous results that stationary values of the $c_i$ correspond to local maxima of $W$, and that $W$ increases monotonically with time. Remarkably, neither of these conclusions is necessarily true, as we now demonstrate.

Maximum values of $W$ occur where $\partial W/\partial c_i = 0$, $(i = 1,2,3,4)$, subject to $\sum c_i = 1$; these conditions are satisfied when

$$W = \tfrac{1}{2}\partial W/\partial c_i, \qquad (i = 1,2,3,4). \tag{8.19}$$

On the other hand, stationary points satisfy (8.15)–(8.18) if $c_i'$ is replaced by $c_i$. The equations so formed may be written

$$W = \tfrac{1}{2}\partial W/\partial c_i \pm c_i^{-1}Rw_{22}(c_2c_3 - c_1c_4), \tag{8.20}$$

where '$+$' applies for $i = 1,4$ and '$-$' for $i = 2,3$. Now eqns. (8.19) and (8.20) are incompatible unless (8.9) holds, and since, as we shall show, this is not necessarily (and indeed is seldom) the case, it follows that points of maximum fitness and equilibrium points are not necessarily identical. It follows that if $c_i = c_i^0$, where $\{c_i^0\}$ is a local maximum of $W$ for which $c_1^0 c_4^0 \neq c_2^0 c_3^0$, then the mean fitness of the population in the following generation will normally decrease; this will certainly be the case if $\{c_i^0\}$ is a global maximum of $W$.

It may be argued that since $c_1(t)c_4(t) - c_2(t)c_3(t) \to 0$ when $w_{ij} = 1$, so that eqn. (8.9) can normally be expected to hold, then the same will happen when the $w_{ij}$ are not all equal, in which case stationary points and points of maximum fitness would coincide. Unfortunately, this argument is not true, as the following numerical example (c.f. Moran (1964)) demonstrates. Suppose that the fitness matrix (8.13) is

$$\begin{pmatrix} 1 & 1 & 0 \\ 1 & 3 & 1 \\ 0 & 1 & 1 \end{pmatrix} \tag{8.21}$$

and let $R = 0{\cdot}5$. With this matrix, $W$ is

$$W = c_1^2 + 2c_1c_2 + 2c_1c_3 + 6c_1c_4 + 6c_2c_3 + c_2c_4 + 2c_3c_4 + c_4^2, \tag{8.22}$$

and the solution of $\partial W/\partial c_i = 0$ (subject to $\sum c_i = 1$) is

$$c_1 = c_4 = \tfrac{1}{6}, \qquad c_2 = c_3 = \tfrac{1}{3}. \qquad (8.23)$$

This, however, is a point of minimum fitness; at it $W = 1\cdot3333$. It is not difficult to show that the absolute maximum of $W$ occurs at the 'boundary' point $c_1 = c_4 = \tfrac{1}{2}$, $c_2 = c_3 = 0$; at this point $W = 2$.

Since, at the point (8.23), eqn. (8.20) does not hold, this point is not a point of equilibrium. The recurrence relations (8.15)–(8.18) show that gametic frequencies in a population having initial frequencies (8.23) take the following values:

| Generation | $c_1 = c_4$ | $c_2 = c_3$ | $W$ |
|---|---|---|---|
| 0 | 0·16666 | 0·33333 | 1·33333 |
| 1 | 0·26042 | 0·23958 | 1·38608 |
| 2 | 0·27555 | 0·22445 | 1·40448 |
| 3 | 0·27703 | 0·22297 | 1·40642 |

The fact that $W$ increases monotonically could have been expected from the fact that (8.23) is a non-equilibrium point of minimum fitness. On the other hand, if the initial frequencies are $c_1 = c_4 = 0\cdot3$, $c_2 = c_3 = 0\cdot2$, gametic frequencies then take the following values:

| Generation | $c_1 = c_4$ | $c_2 = c_3$ | $W$ |
|---|---|---|---|
| 0 | 0·30000 | 0·20000 | 1·44000 |
| 1 | 0·28125 | 0·21875 | 1·41211 |
| 2 | 0·27801 | 0·22199 | 1·40772 |
| 3 | 0·27745 | 0·22255 | 1·40698 |

In this case $W$ decreases monotonically. This remarkable result runs against intuition and against the results for single loci found in Chapter 3. So far as the equilibrium value $c$ of $c_1$ is concerned, eqn. (8.15), together with considerations of symmetry, shows that $c$ satisfies the equation

$$6c^3 - 4c^2 + 2c - \tfrac{3}{8} = 0,$$

the relevant solution of which is $c = 0\cdot27734 \ldots$ . At this point $W = 1\cdot40684$, and this is clearly the equilibrium towards which both sequences above are tending. At this point, $c_1 c_4 - c_2 c_3 = 0\cdot02736$, showing that (8.12) is not true. Since we have been assuming unlinked loci, it is clear that the term 'coefficient of linkage disequilibrium' is misleading inasmuch as it implies that non-zero values of $c_1 c_4 - c_2 c_3$

are due to linkage. Since we have seen that, irrespective of the degree of linkage, $c_1c_4 - c_2c_3$ can normally be taken as zero when there are no selective differences, and since, even for unlinked loci, $c_1c_4 - c_2c_3$ is not normally zero when selective differences exist, a better term for the quantity $c_1c_4 - c_2c_3$ is 'coefficient of epistatic disequilibrium'.

The not necessarily increasing nature of $W$ raises a number of questions of some mathematical interest. It may be asked, for example, whether conditions can be imposed on the matrix (8.13) which ensure that $W$ increases. A second approach is to abandon consideration of $W$ and to ask whether there exists some other function $Y$ satisfying the following three conditions: (i) $Y$ is non-decreasing, (ii) $Y$ is maximized at stationary points, (iii) $Y$ has some genetical meaning as an overall measure of fitness. Little progress has been made in this direction; indeed it seems very difficult to find a function satisfying even two of the conditions.

Genetically, one may ask whether the phenomenon of decreasing $W$ will occur frequently, or whether the numerical example above is atypical. That $W$ should usually increase is suggested by the following argument, due to Kimura (1965). It has been shown above that when there are no selective differences, the function $K$, defined by

$$K = c_1c_4/c_2c_3, \qquad (8.24)$$

will approach unity. It is therefore reasonable to consider changes in $K$ in the more general case where selection is allowed. To a close approximation,

$$\Delta K = K[c_1^{-1}\Delta c_1 - c_2^{-1}\Delta c_2 - c_3^{-1}\Delta c_3 + c_4^{-1}\Delta c_4]. \qquad (8.25)$$

The recurrence relations (8.15) − (8.18) can be used to reduce this equation to

$$W\Delta K = KD + w_{22}R(1-K)(c_1 + c_4 + Kc_2 + Kc_3), \qquad (8.26)$$

where

$$D = w_{11}c_1 + w_{12}(c_2 - c_1) + w_{21}(c_3 - c_1) + w_{22}(c_4 - c_3 - c_2 + c_1)$$
$$+ w_{23}(c_2 - c_4) - w_{31}c_3 + w_{32}(c_3 - c_4) + w_{33}c_4. \qquad (8.27)$$

Kimura now argues that since $D$ will usually be small, eqn. (8.26) shows that if $K$ exceeds its equilibrium value near $K = 1$, $\Delta K$ will be negative, while if $K < 1$, $\Delta K$ is positive. Thus $K$ should rapidly settle down to its equilibrium value and remain practically constant thereafter.

However, when $\Delta K$ is identically zero, it can be shown that

$$\Delta W = V, \tag{8.28}$$

where $V$ is the 'additive' component of the genetic variance (defined for two loci in a manner extending that used in Chapter 3). In this case, not only should $\Delta W$ be positive, but indeed $\Delta W$ should have an interpretation as an essential component of genetic variance. Kimura gives a number of numerical examples which suggest that both these conclusions are usually valid.

While this analysis is very appealing, the approximations involved ensure that the suggested conclusion is not always true. Kimura supplies one example where $K \to \infty$, and a second for which $K \to 1$ and yet $\Delta W$ is always negative. Generally it appears that Kimura's analysis is most likely to apply when the $w_{ij}$ are almost equal and $R$ is not too small.

## 8.4. What is 'interaction'?

The complex behaviour of populations whose fitness depends on two loci leads to one to ask whether the complexity is removed if the two loci concerned do not 'interact'. To answer this question, it must first of all be decided what is meant by 'interaction' between two loci. In this section a discussion of this problem will be given; our analysis will follow that of Moran (1965).

The two most obvious ways of saying that the loci 'A' and 'B' do not interact are, firstly, to suppose that we can find parameters $u_i$ and $v_j$ such that $w_{ij} = u_i + v_j$, and secondly that such parameters can be found so that $w_{ij} = u_i v_j$. These will be called the 'additive' and 'multiplicative' cases respectively; it is reasonable to expect that if in some sense the '$A$' and '$B$' loci are to act independently, then the fitness parameters will assume one or other of these forms.

The case of greatest interest is when $u_2$ exceeds both $u_1$ and $u_3$ and $v_2$ exceeds $v_1$ and $v_3$; if this is not so, there is no 'internal' equilibrium and the problem degenerates. It is not difficult to see that in both additive and multiplicative cases, the point

$$c_1 = Pp, \quad c_2 = Pq, \quad c_3 = Qp, \quad c_4 = Qq \tag{8.29}$$

is a point of equilibrium, where

$$P = 1-Q = (u_2-u_3)/(2u_2-u_1-u_3), \tag{8.30}$$

$$p = 1-q = (v_2-v_3)/(2v_2-v_1-v_3). \tag{8.31}$$

That the point (8.29) is one of equilibrium is to be expected from our knowledge of single locus behaviour; to this extent the two loci can be thought of as acting independently. Further, the point (8.29) has the additional property that at it (8.9) is satisfied; one would reasonably hope that this would be the case for 'independent' loci. It is not difficult to show that, in the additive case, the point (8.29) is a point of stable equilibrium and is the only internal equilibrium point when $R > 0$, which we assume to be the case; the corresponding statement is not true in the multiplicative case. These results make one suspect that the additive case may well correspond to 'independent' loci. While indeed it appears that this case is the closest that one can get to independence, this suspicion is not completely justified since it is not always true that $c_1 c_4 - c_2 c_3$ decreases in absolute value from one generation to the next, nor is it always true that the gene frequencies $c_1 + c_2$, $c_1 + c_3$ converge monotonically to their final values.

It is useful to discuss the multiplicative case by first considering a particular example, and we therefore suppose that the fitness matrix (8.13) is of the form

$$\begin{pmatrix} 1 & 1\cdot1 & 1 \\ 1\cdot1 & a & 1.1 \\ 1 & 1\cdot1 & 1 \end{pmatrix} \qquad (8.32)$$

Here $a = 1\cdot2$ is 'additive' and $a = 1\cdot21$ is 'multiplicative'. Equations (8.15)–(8.18) become

$$W c_1' = c_1{}^2 + 1\cdot1 c_1 c_2 + 1\cdot1 c_1 c_3 + a c_1 c_4 + Ra(c_2 c_3 - c_1 c_4), \quad (8.33)$$

$$W c_2' = 1\cdot1 c_1 c_2 + c_2{}^2 + a c_2 c_3 + 1\cdot1 c_2 c_4 - Ra(c_2 c_3 - c_1 c_4), \quad (8.34)$$

$$W c_3' = 1\cdot1 c_1 c_3 + a c_2 c_3 + c_3{}^2 + 1\cdot1 c_3 c_4 - Ra(c_2 c_3 - c_1 c_4), \quad (8.35)$$

$$W c_4' = a c_1 c_4 + 1\cdot1 c_2 c_4 + 1\cdot1 c_3 c_4 + c_4{}^2 + Ra(c_2 c_3 - c_1 c_4), \quad (8.36)$$

where

$$W = 1 + 0\cdot2(c_1 c_2 + c_1 c_3 + c_2 c_4 + c_3 c_4) + 2(a-1)(c_1 c_4 + c_2 c_3). \qquad (8.37)$$

From symmetry, if $c_1 = c_4$ and $c_2 = c_3$, then $c_1' = c_4'$, $c_2' = c_3'$, while $c_1' + c_2' = \frac{1}{2}$. Equations (8.33) and (8.34) then yield

$$c_1' - c_2' = F(c_1 - c_2), \qquad (8.38)$$

where

$$F = (1 + a - 2Ra)/\{1 + a + (9\cdot6 - 8a)c_1 c_2\}. \qquad (8.39)$$

In the additive case, $F$ takes the value $1 - \{2Ra/(1+a)\}$. Clearly this always lies in $(0,1)$, so that $c_1 - c_2 \to 0$. Thus each $c_i \to \frac{1}{4}$ which is, as claimed above, a point of stable equilibrium. When $a > 1\cdot2$ things are not so simple. If each $c_i$ is approximately $\frac{1}{4}$, then to a close approximation $F$ will exceed unity when

$$R < (a-1\cdot2)/4a. \tag{8.40}$$

In the multiplicative case this condition becomes

$$R < 1/484. \tag{8.41}$$

Thus, with extremely close linkage, the equilibrium $c_i = \frac{1}{4}$ is unstable; in such cases we would hardly describe the two loci as being independent. To take this point even further, eqns. (8.33)-(8.36) show that if (8.9) holds in any given generation, it will hold in all subsequent generations; if however it is not initially zero, the quantity $c_1c_4 - c_2c_3$ can steadily increase in absolute value. This follows because eqns. (8.33)–(8.36) yield

$$(c_2{'}c_3{'} - c_1{'}c_4{'})W^2 = (c_2c_3 - c_1c_4)\,(\Omega - 1\cdot21RW),$$

where

$$\Omega = 1\cdot21c_1{}^2 + 2\cdot431c_1c_2 + 1\cdot21c_2{}^2 + 2\cdot431c_1c_3 + 2\cdot4641c_1c_4$$

$$+ 2\cdot431c_2c_4 + 1\cdot21c_3{}^2 + 2\cdot431c_3c_4 + 1\cdot21c_4{}^2. \tag{8.42}$$

If $\Omega - 1\cdot21RW > W^2$, then clearly

$$|c_2{'}c_3{'} - c_1{'}c_4{'}| > |c_2c_3 - c_1c_4|. \tag{8.43}$$

If, for example, $c_1 = c_4 = \frac{1}{3}$, $c_2 = c_3 = \frac{1}{6}$, then

$$\Omega = 1\cdot21857, \qquad W = 1\cdot10278, \qquad W^2 = 1\cdot21612,$$

so that the inequality (8.43) holds whenever

$$R < 0\cdot00184. \tag{8.44}$$

It is, indeed, possible to show that (8.43) holds for all pairs of consecutive generations. Since, by symmetry, we can put $c_1 = c_4 = c$ (say), $c_2 = c_3 = \frac{1}{2} - c$, eqn.s (8.37) and (8.42) yield

$$W = 1\cdot105 - 0\cdot02c + 0\cdot04c^2, \qquad \Omega = 1\cdot105W.$$

The inequality $\Omega - 1\cdot21RW > W^2$ then becomes

$$0\cdot02c(1-2c) > 1\cdot21R. \tag{8.45}$$

(In the particular case $c = \frac{1}{3}$, this reduces to (8.44).) Since the inequality (8.44) implies the inequality (8.41), the point $c_i = \frac{1}{4}$ is certainly unstable when (8.44) holds. In this case, the two equilibrium values of $c$ can be shown, from (8.33), to be

$$\tfrac{1}{4}[1 + \{1 - 484R\}^{\frac{1}{2}}], \quad \tfrac{1}{4}[1 - \{1 - 484R\}^{\frac{1}{2}}]. \tag{8.46}$$

If initially $c = \frac{1}{3}$, then $c$ will steadily approach the former point, which is stable; furthermore, the condition that this stable equilibrium point exceeds $\frac{1}{3}$ reduces to (8.44), in which case (8.45) and thus (8.43) hold throughout the entire process.

In passing it is interesting to note that for this process,

$$c_1 c_4 / c_2 c_3 \rightarrow [\{1 + (1 - 484\text{R})^{\frac{1}{2}}\} / \{1 - (1 - 484R)^{\frac{1}{2}}\}]^2,$$

so that if $R$ is small, $c_1 c_4 / c_2 c_3$ will approach a large number; this is one of the cases in which linkage is so tight that the analysis of Kimura referred to in the previous section will not hold.

Even the brief examination of independence carried out in this section shows that the concept is more complex than might initially appear. Probably the closest approach to independence is where fitnesses are additive, but even here the behaviour is not completely straightforward and it is probable that conditions for complete independence do not exist.

# CHAPTER 9

# Linkage

It has been shown in the previous chapter that the degree of linkage between two loci can affect quite strongly the evolutionary behaviour of populations whose fitness depends on these loci. In this chapter a more systematic examination of the various evolutionary effects that such linkage can have will be made. It might be remarked before starting that this cannot hope to be anything but a superficial examination; for important facets of linkage behaviour not covered in the following sections, see Felsenstein (1965), Haldane (1963), Jain and Allard (1965), Karlin, McGregor and Bodmer (1966), Kojima and Schaffer (1967), Levins (1965), Parsons (1963) and Turner (1967).

## 9.1. Equilibrium points and their stability behaviour

The first problem to be considered concerns the position and nature of non-trivial stationary points of processes for which gamete frequencies in successive generations are given by the relations (8.15)–(8.18), which may be rewritten for convenience

$$c_i' = \phi_i(c_1 c_2, c_3, c_4), \qquad i = 1, \ldots, 4. \tag{9.1}$$

If the frequencies $c_i = c_i^*$ are stationary, it follows that

$$c_i^* = \phi_i(c_1^*, c_2^*, c_3^*, c_4^*), \qquad i = 1, \ldots, 4. \tag{9.2}$$

The solution of these equations may well be rather difficult to find, and generally it appears that numerical methods must be used. There are, however, various special cases where eqn. (9.2) can be solved explicitly; these will be considered later. For the moment, it is sufficient to note the obvious fact that the position of the points $c_i^*$ will depend, to some extent, on the value of the recombination fraction $R$.

A more important question is to describe the stability behaviour of any equilibrium point, since points of unstable equilibrium are of little interest. The stability nature of (9.2) is tested in the usual

H

way by supposing that each $c_i$ differs from its equilibrium value $c_i{}^*$ by a small amount $\delta_i$. Then

$$c_i{}^* + \delta_i' = \phi_i(c_1{}^* + \delta_1, c_2{}^* + \delta_2, c_3{}^* + \delta_3, c_4{}^* + \delta_4)$$

$$= \phi_i(c_1{}^*, c_2{}^*, c_3{}^*, c_4{}^*) + \sum_i \delta_j \left[\frac{\partial \phi_i}{\partial c_j}\right]^* + 0(\delta_j{}^2), \ i = 1, \ldots, 4. \quad (9.3)$$

Here the notation $[f]^*$ means the value of the function $f$ evaluated at the equilibrium point. Ignoring small-order terms, this becomes

$$\boldsymbol{\delta}' = A\boldsymbol{\delta}, \quad (9.4)$$

where the $i-j^{\text{th}}$ term in $A$ is $\left[\dfrac{\partial \phi_i}{\partial c_j}\right]^*$. It is clear that

$$\boldsymbol{\delta}^{(n)} = A^n \boldsymbol{\delta},$$

so that the necessary and sufficient condition that $\boldsymbol{\delta}^{(n)} \to 0$, implying stability of $c^*$, is that all eigenvalues of $A$ be less than unity in absolute value. Writing more fully

$$\phi_i(c_1, c_2, c_3, c_4) = W^{-1} h_i(c_1, c_2, c_3, c_4),$$

we have

$$a_{ij} = \left[\frac{\partial \phi_i}{\partial c_j}\right]^* = (W^*)^{-1} \left[\left\{\frac{\partial h_i}{\partial c_j}\right\}^* - c_i{}^* \left\{\frac{\partial W}{\partial c_j}\right\}^*\right]. \quad (9.5)$$

A further simplification is possible by noting the identity $\sum \delta_i = 0$, from which it follows that

$$\begin{pmatrix} \delta_1' \\ \delta_2' \\ \delta_3' \end{pmatrix} = B \begin{pmatrix} \delta_1 \\ \delta_2 \\ \delta_3 \end{pmatrix}, \quad (9.6)$$

where the typical element $b_{ij}$ of $B$ is

$$b_{ij} = a_{ij} - a_{i4}. \quad (9.7)$$

The necessary and sufficient condition for $c^*$ to be stable is then that all eigenvalues of $B$ be less than unity in absolute value; despite loss of symmetry, this condition is usually easier to apply than the corresponding condition using $A$. We now present several examples.

*Example* 1. (*Kimura*, (1956))

Suppose that the matrix (8.13) is of the form

$$\begin{pmatrix} 1+s & 1+t & 1-s \\ 1 & 1+t & 1 \\ 1-s & 1+t & 1+s \end{pmatrix} \qquad (9.8)$$

where $0 < s < t$. From symmetry, any internal equilibrium frequency $c_1^*$ must equal $c_4^*$, and similarly $c_2^* = c_3^*$. Writing $c_1^* = c_4^* = c$, $c_2^* = c_3^* = \frac{1}{2} - c$, eqn. (8.15) becomes

$$sc^2 + (R + Rt - \tfrac{1}{2}s)c - \tfrac{1}{4}R(1+t) = 0. \qquad (9.9)$$

It is easy to show that the only relevant solution of eqn. (9.9) lies in $(\frac{1}{4}, \frac{1}{2})$, and is, explicitly,

$$c = [s - 2R(1+t) + \{s^2 + 4R^2(1+t)^2\}^{\frac{1}{2}}]/4s. \qquad (9.10)$$

The value of $R$ has an appreciable influence on this solution. For $R$ very small, the solution is approximately $c = \frac{1}{2}$, in which case nearly all individuals are $A_1 A_1 B_1 B_1$, $A_1 B_1 / A_2 B_2$ or $A_2 A_2 B_2 B_2$. Here the equilibrium is similar to that at a single locus where both homozygotes have fitness $1+s$ and the heterozygote has fitness $1+t$.

The next problem is to investigate the stability behaviour of (9.10). After a considerable algebra we find that the elements $b_{ij}$ of $B$ are

$$b_{11} = W^{-1}(1 + 2sc + \tfrac{1}{2}t - ct), \; b_{12} = W^{-1}\{sc + \tfrac{1}{2}R(1+t)\},$$

$$b_{13} = W^{-1}\{sc - ct + \tfrac{1}{2}R(1+t)\}, \; b_{21} = W^{-1}\{t(\tfrac{1}{2} - c)\},$$

$$b_{22} = W^{-1}\{1 - s(\tfrac{1}{2} - c) + \tfrac{1}{2}t - \tfrac{1}{2}R(1+t)\},$$

$$b_{23} = W^{-1}\{\tfrac{1}{2}t - ct - \tfrac{1}{2}R(1+t) + \tfrac{1}{2}s - sc\},$$

$$b_{31} = W^{-1}\{-t(\tfrac{1}{2} - c)\}, \; b_{32} = W^{-1}\{(\tfrac{1}{2} - c)s - \tfrac{1}{2}R(1+t)\},$$

$$b_{33} = W^{-1}\{1 + ct - \tfrac{1}{2}R(1+t) - s(\tfrac{1}{2} - c)\},$$

where

$$W = 1 + \tfrac{1}{2}t + s(2c - \tfrac{1}{2}).$$

If $\psi = W(\lambda - 1)$, the eigenvalue equation $|B - \lambda I| = 0$ becomes

$$\psi^3 + \tfrac{1}{2}(t+v)\psi^2 + [\tfrac{1}{2}t\{v - R(1+t)\} - \tfrac{1}{4}s^2]\psi - \tfrac{1}{8}v\{s^2 - tv + 2Rt(1+t)\} = 0, \qquad (9.11)$$

where $v = \{s^2 + 4R^2(1+t)^2\}^{\frac{1}{2}}$. By inspection it follows that $\psi = -\frac{1}{2}v$ $= \psi_1$ is a solution from (9.11); from this it follows that the remaining two eigenvalues are

$$\psi = \psi_2 = \tfrac{1}{4}[-t + \{t^2 + 4s^2 - 4tv + 8tR(1+t)\}^{\frac{1}{2}}], \tag{9.12}$$

$$\psi = \psi_3 = \tfrac{1}{4}[-t - \{t^2 + 4s^2 - 4tv + 8tR(1+t)\}^{\frac{1}{2}}]. \tag{9.13}$$

The eigenvalue $\psi_1$ is clearly real; the eigenvalues $\psi_2$ and $\psi_3$ will be real provided

$$t^2 + 4s^2 - 4tv + 8tR(1+t) \geq 0.$$

The inequality $v < s + 2R(1+t)$ shows that this is always the case; all eigenvalues are therefore real. The condition $-1 < \lambda < 1$ then becomes $-2W < \psi < 0$. The inequality $-2W < \psi_i (i = 1,2,3)$ is easily proved; the inequalities $\psi_1 < 0$, $\psi_3 < 0$ always hold, while $\psi_2$ is negative whenever

$$R < (t^2 - s^2)/4t(1+t). \tag{9.14}$$

Thus (9.10) is stable if and only if the $A$ and $B$ loci are sufficiently closely linked, the limit for the recombination fraction being given by (9.14). Thus linkage has a strong qualitative, as well as quantitative, effect on the stability behaviour of the population.

*Example 2. (Lewontin and Kojima, (1960), Lewontin (1964a)).*

When the fitness matrix is of the form

$$\begin{pmatrix} 1-s & 1-t & 1-s \\ 1-u & 1 & 1-u \\ 1-s & 1-t & 1-s \end{pmatrix} \tag{9.15}$$

symmetry again shows that $c_1{}^* = c_4{}^* = \frac{1}{2} - c_2{}^* = \frac{1}{2} - c_3{}^* = c$. Equation (8.15) becomes

$$(s-t-u)\,(4c^3 - 3c^2 + \tfrac{1}{2}c) - R(c - \tfrac{1}{4}) = 0,$$

the solutions of which are

$$c = \tfrac{1}{4}, \quad c = \tfrac{1}{4} \pm \tfrac{1}{4}\{1 - 4R/(t+u-s)\}^{\frac{1}{2}}, \tag{9.16}$$

the latter solutions existing only when

$$R \leq \tfrac{1}{4}(t+u-s). \tag{9.17}$$

Of these we need examine in detail only the solutions

$$c = \tfrac{1}{4}, \quad c = c_0 = \tfrac{1}{4} + \tfrac{1}{4}\{1 - 4R/(t+u-s)\}^{\frac{1}{2}}. \tag{9.18}$$

The mean fitnesses of the population at the equilibrium points $c = \frac{1}{4}$, $c = c_0$ are

$$W = W_1 = 1 - \tfrac{1}{4}(t-u+s), \quad W = W_0 = 1 - \tfrac{1}{2}s - R. \quad (9.19)$$

The condition that $c = \frac{1}{4}$ be stable is found to be

$$R \geqq \tfrac{1}{4}(t+u-s) \quad (9.20)$$

and

$$s > |t-u|. \quad (9.21)$$

Again after considerable algebraic manipulation, (c.f. Ewens, (1968a)), it may be shown that $c_0$ is stable whenever

$$4R^2(t+u-s)+2R\{2s^2-u^2-t^2-s(t+u)\}+s(t+u-s)^2 > 0. \quad (9.22)$$

This condition always holds near $R = 0$, so that $c_0$ is always stable (if it exists) if linkage is extremely tight. At the other extreme, when $R = \frac{1}{4}(t+u-s)$, (9.22) becomes

$$|t-u| < s. \quad (9.23)$$

Thus if the solutions of the equation

$$4R^2(t+u-s)+2R\{2s^2-u^2-t^2-s(t+u)\}+s(t+u-s)^2 = 0$$

are denoted by $R_1$ and $R_2$, $(R_1 < R_2)$ the region of stability of $c_0$ can be of three different forms:

(i) of the form $(0, R_1)$, (when $R_1 > 0$ but (9.23) does not hold), (9.24)

(ii) of the form $\{0, \frac{1}{4}(t+u-s)\}$, (when $R_1 > \frac{1}{4}(t+u-s)$),

(iii) of the form

$$[(0, R_1) \cup (R_2, \tfrac{1}{4}(t+u-s))], \quad (9.25)$$

(when $0 < R_1 < R_2 < \frac{1}{4}(t+u-s)$).

The following three numerical examples exemplify each of these:

(i) $t = 0 \cdot 7, u = 0 \cdot 4, s = 0 \cdot 1$. Here $c_0$ exists whenever $R < 0 \cdot 25$, but is stable only when $R < 0 \cdot 09$.

(ii) $t = 0 \cdot 18, u = 0 \cdot 22, s = 0 \cdot 35$. Here $R_1 < R_2 < 0$ and $c_0$ is stable wherever defined.

(iii) $t = 0 \cdot 78, u = 0 \cdot 82, s = 0 \cdot 1$. Here the stability region is
$$(0, 0 \cdot 1005) \cup (0 \cdot 3731, 0 \cdot 375).$$

105

It is remarkable, and perhaps unexpected, that stability regions such as (9.25) can occur; while a verbal discussion of such behaviour will be deferred, it might be remarked generally at this stage that regions of the form (9.25) exist whenever $t$ and $u$ are jointly sufficiently large compared to $s$.

More explicit conditions may be found when $t = u$, in which case (9.23) holds automatically. The stability region is therefore either of the form (9.24) or of the form (9.25), and (9.22) shows that the latter applies whenever

$$s/t < 1 - (1/\sqrt{(2)}).\qquad(9.26)$$

As a numerical example, when $s = 0\cdot1$, $t = u = 0\cdot35$, condition (9.26) holds, and $c_0$ is stable when

$$0 < R < 0\cdot1125 \quad \text{or} \quad 0\cdot1333 < R < 0\cdot15.$$

The nature of this region makes clear the extraordinary effect of linkage in this model.

*Example 3. Multiplicative fitnesses*

In the previous chapter it was shown that when $w_{ij}$ could be expressed in the form $u_i v_j$, then (8.29) is a point of equilibrium which, in the numerical case (8.32) with $a = 1\cdot21$, is stable only when linkage is sufficiently loose. This appears to be typical of a general result: if linkage is loose, (8.29) is stable; if linkage is tight, a stable equilibrium analogous to (8.46) is achieved. Thus when $u_1 = u_3 = 1 - a$, $v_1 = v_2 = 1 - b$, the equilibrium value $c$ of $c_1$ satisfies the equation

$$4abc^3 - 3abc^2 + \tfrac{1}{2}abc + R(c - \tfrac{1}{4}) = 0,$$

so that

$$c = \tfrac{1}{4}, \quad \tfrac{1}{4} \pm \tfrac{1}{4}\{1 - 4R/ab\}^{\frac{1}{2}}.\qquad(9.27)$$

The latter two solutions exist only when $R < \tfrac{1}{4}ab$, in which case it is not difficult to show that they are stable; otherwise, $c = \tfrac{1}{4}$ is stable. In the more general case when $u_1 \neq u_3$, $v_1 \neq v_3$, it appears impossible to obtain such explicit expressions, and resort must be made to numerical simulation.

### 9.2. Increase in frequency of rare genes

It is interesting to consider the conditions under which, if nearly all individuals initially are $- - B_2 B_2$, the frequency of $B_1$ will increase.

This can be done by using eqns. (8.15) and (8.16), provided that some extremely reasonable simplifications are made. Thus since we are interested in the increase in $B_1$ genes when $B_1$ is rare, the existence of $B_1B_1$ individuals can be ignored. It may further be supposed that, when the process starts, the frequencies of $A_1A_1B_2B_2$, $A_1A_2B_2B_2$ and $A_2A_2B_2B_2$ are in Hardy-Weinberg form $p^2, 2pq, q^2$. Finally, it may be supposed that throughout the period in which we are interested, $c_3 = p, c_4 = q$, and that

$$W = p^2w_{31} + 2pqw_{32} + q^2w_{33}. \tag{9.28}$$

The recurrence relations (8.15) and (8.16) now become

$$c_1' = W^{-1}[\{w_{21}p + w_{22}(1-R)q\}c_1 + Rw_{22}pc_2], \tag{9.29}$$

$$c_2' = W^{-1}[Rw_{22}qc_1 + \{w_{22}p(1-R) + w_{23}q\}c_2], \tag{9.30}$$

that is to say,

$$c' = Tc, \tag{9.31}$$

where

$$T = W^{-1}\begin{pmatrix} W_1 - Rw_{22}q & Rw_{22}p \\ Rw_{22}q & W_2 - Rw_{22}p \end{pmatrix} \tag{9.32}$$

and

$$W_1 = w_{21}p + w_{22}q, \quad W_2 = w_{22}p + w_{23}q. \tag{9.33}$$

The frequency of $B_1$ will thus increase eventually if and only if the larger eigenvalue of $T$ exceeds unity. Now the eigenvalues of $T$ are

$$\lambda_1 = (2W)^{-1}[W_1 + W_2 - Rw_{22} + \{(W_1 - W_2)^2$$
$$+ 2Rw_{22}(p-q)(W_1 - W_2) + R^2w_{22}^2\}^{\frac{1}{2}}], \tag{9.34}$$

$$\lambda_2 = (2W)^{-1}[W_1 + W_2 - Rw_{22} - \{(W_1 - W_2)^2$$
$$+ 2Rw_{22}(p-q)(W_1 - W_2) + R^2w_{22}^2\}^{\frac{1}{2}}]. \tag{9.35}$$

Both eigenvalues decrease with $R$ for $R$ in $(0, \frac{1}{2})$, except in the special case $W_1 = W_2$, when $\lambda_1$ is constant and $\lambda_2$ decreases with $R$. Thus, apart from this particular case, the condition $\lambda_1 > 1$ must hold for all $R$ in $(0, \frac{1}{2})$, for no $R$ in $(0, \frac{1}{2})$, or for values of $R$ in a region of the form $R < R_0$, where $0 < R_0 < \frac{1}{2}$. The first applies when both $W_1$ and $W_2$ exceed $W$, the second when $W$ exceeds both $W_1$ and $W_2$, while if $W$ lies between $W_1$ and $W_2$, $\lambda_1$ exceeds unity only when

$$Rw_{22}(W - W^*) < -(W - W_1)(W - W_2), \tag{9.36}$$

where

$$W^* = p^2w_{21}+2pqw_{22}+q^2w_{23} = pW_1+qW_2. \qquad (9.37)$$

This requirement holds automatically when $W^*$ exceeds $W$, but otherwise holds only for $R$ sufficiently small.

It is worth while considering some of the implications of these results, and to do this we first consider $W^*$, $W_1$ and $W_2$. The quantity $W_1$ can loosely be described as the fitness of the gamete $A_1B_1$, since with probability $p$ such a gamete combines with an $A_1B_2$ gamete to form a zygote having fitness $w_{21}$, while with probability $q$ it combines with an $A_2B_2$ gamete to form a zygote having fitness $w_{22}$. Similarly, $W_2$ can be called the fitness of $A_2B_1$ gametes. Finally, $W^*$ is the mean fitness of $B_1B_2$ individuals. The following conclusions can then be drawn: (i) if both $A_1B_1$ and $A_2B_1$ gametes have fitness higher than that of the mean fitness of the population before $B_1$ is introduced, then the frequency of $B_1$ will increase; (ii) if, on the other hand, the fitnesses of both $A_1B_1$ and $A_2B_1$ gametes are less than this, the frequency of $B_1$ will decrease; (iii) if the mean fitness of $B_1B_2$ exceeds that of $B_2B_2$, then the frequency of $B_1$ will increase, (this subsumes (1)), and (iv) even if the mean fitness of $B_1B_2$ does not exceed that of $B_2B_2$, the frequency of $B_1$ can increase, provided that linkage between $A$ and $B$ loci is sufficiently close, the limit to the recombination fraction being given by (9.36).

Statements (i), (ii) and (iii) agree with intuition. The really startling result is (iv); linkage can cause a 'less fit' gene to replace steadily a 'more fit' one, at least when the former is rare. These general results are now illustrated by some particular examples.

*Example* 1.

If, for the fitness matrix (9.8), nearly all individuals are $B_2B_2$, the frequency $p$ of $A_1$ will be $(t-s)(2t)^{-1}$, and

$$W = (2t+t^2+s^2)(2t)^{-1}, \quad W^* = (2t+t^2-s^2)(2t)^{-1}, \quad (9.38)$$

$$W_1 = (2t+t^2+st)(2t)^{-1}, \quad W_2 = (2t+t^2+s^2)(2t)^{-1}.$$

Since $W_2 < W^* < W < W_1$, the frequency of $B_1$ will increase only if linkage between $A$ and $B$ loci is sufficiently tight: specifically, (9.36) shows that this happens when

$$R < (t^2-s^2)/4t(1+t).$$

This condition is identical to that which ensures the existence of a stable internal equilibrium point, and thus a complete analysis of the behaviour of gamete frequencies for this model can easily be made.

*Example 2*

Kimura (1956) has proposed a model to explain the polymorphism at the two linked loci in the snail *cepea nemoralis* which control respectively colour and shell banding. A simplified model of the fitnesses of the various genotypes, together with their phenotypic expressions, is as follows.

|  | *Pink* | | *Yellow* |
|---|---|---|---|
|  | $A_1A_1$ | $A_1A_2$ | $A_2A_2$ |
| *Unbanded* | | | |
| $B_1B_1$ | $1+s$ | $1+s+t$ | $1-s$ |
| $B_1B_2$ | $1+s$ | $1+s+t$ | $1-s$ |
| *Banded* | | | |
| $B_2B_2$ | $1-s$ | $1-s+t$ | $1+s$ |

$$(9.39)$$

If $0 < s < \frac{1}{2}t$, there will be a polymorphism $A_1 - A_2$ in the absence of $B_1$, with $p = $ frequency of $A_1 = (t-2s)/(2t-2s)$. The inequality (9.36) becomes

$$R(4st - 2s^2 - t^2) < 4s^2(t-s)/(1+s+t), \qquad (9.40)$$

so that the frequency of 'unbanded' will increase when (9.40) holds. A sufficient, but not necessary, condition for this to happen is

$$2s^2 + t^2 > 4st, \qquad (9.41)$$

a condition which can be written more meaningfully (c.f. Bodmer and Parsons, (1962)) as $s/t < 1 - (1/\sqrt{(2)})$. In general terms, it is clear why this behaviour occurs. Condition (9.41) is equivalent to the condition $W^* > W$, and we know that this is sufficient for increase in frequency of $B_1$; otherwise $B_1$ increases only when linkage between 'colour' and 'banding' loci is sufficiently close.

## 9.3. Initial increase in two mutant genes

A situation rather similar to that in the previous section is when (say) nearly all individuals are initially $A_2A_2B_2B_2$, except for a small number of individuals carrying an $A_1$ or $B_1$ gene. Here we are interested in the circumstances under which the frequencies of both $A_1$ and $B_1$ can increase.

It is reasonably clear that the frequency of $A_1$ will increase when $w_{32} > w_{33}$, while that of $B_1$ will increase when $w_{23} > w_{33}$. Neither of these conditions, however, is necessary for the frequencies of both $A_1$ and $B_1$ to increase, for if $w_{22} > w_{33}$ and linkage is sufficiently tight, the frequency of $A_1B_1$ gametes increases irrespective of the values of $w_{23}$ and $w_{32}$. A more precise condition is found by replacing, as is reasonable, $W$ by $w_{33}$ in eqn. (9.29) and at the same time putting $p = 0$. This gives

$$c_1' = w_{33}^{-1}[c_1 w_{22}(1 - R) + R w_{22} c_2 c_3], \qquad (9.42)$$

so that

$$c_1' \geqq c_1[w_{22} w_{33}^{-1}(1 - R)]. \qquad (9.43)$$

It follows that $c_1$ steadily increases in frequency when

$$R < w_{22}^{-1}(w_{22} - w_{33}). \qquad (9.44)$$

This condition is sufficient but not necessary for the frequencies of $A_1$ and $A_2$ to increase; necessary conditions seem to be very hard to find, the only general result being that when $w_{23} < w_{33}$, $w_{32} < w_{33}$, the condition (9.44) is probably very close to being necessary as well as sufficient. In any event, it is clear from (9.44) that the degree of linkage between the loci at which they occur plays an important part in determining whether or not two rare genes increase in frequency.

## 9.4. Sudden changes in mean fitness

It has been observed in many experiments in which the fitness of an individual depends on his genetic consitution at two or more loci, that during selection more or less sudden changes sometimes occur in the mean population fitness. Examples of this phenomenon date to the experiments of Mather (1941), who carried out experiments in which selection was made, in various *Drosophila* populations, for individuals having high chaetae number. A noticeable feature of these experiments in some cases was that after a rather steady mean chaetae number had persisted for some generations a sudden increase in mean number, which in this context can be identified with a sudden increase in mean fitness, occurred, after which the mean number remained steady at the higher value. This sudden increase was attributed to recombination and the aim of this section is to show, by considering several numerical examples, that this is a most reasonable hypothesis. It should, however, be pointed out that the numerical values used for fitnesses in these examples are not intended to correspond to those in Mather's experiments; they are intended to show that the phenomenon considered is possibly rather general.

If the fitness matrix (8.13) is

$$\begin{pmatrix} 1 & 1 & \frac{1}{2} \\ 1 & \frac{3}{4} & \frac{1}{2} \\ 1 & \frac{1}{2} & \frac{1}{2} \end{pmatrix} \tag{9.45}$$

and if initially $c_1 = 0 \cdot 0001$, $c_2 = c_3 = 0 \cdot 0099$, $c_4 = 0 \cdot 9801$, then the frequencies $c_1$, $c_2$ and $c_3$ will (initially) increase at the expense of $c_4$. A summary of the values assumed by these frequencies in various generations, together with the mean population fitness, is given in Table 9.1 for both tightly linked and unlinked loci.

TABLE 9.1. Frequencies of gametes $A_1B_1$, $A_2B_1$, $A_1B_2$, and $A_2B_2$ in various generations in the model (9.45), together with mean population fitness.

| Generation | Frequency of | | | | Mean fitness |
| | $A_1B_1$ | $A_2B_1$ | $A_1B_2$ | $A_2B_2$ | of population |
|---|---|---|---|---|---|
| | | | $R = \cdot01$ | | |
| 0 | 0·0001 | 0·0099 | 0·0099 | 0·9801 | 0·5001 |
| 5 | 0·0008 | 0·0102 | 0·0107 | 0·9784 | 0·5004 |
| 10 | 0·0057 | 0·0106 | 0·0117 | 0·9720 | 0·5030 |
| 15 | 0·0409 | 0·0119 | 0·0138 | 0·9335 | 0·5211 |
| 20 | 0·2287 | 0·0176 | 0·0206 | 0·7331 | 0·6191 |
| 25 | 0·6187 | 0·0279 | 0·0330 | 0·3204 | 0·8292 |
| 30 | 0·8479 | 0·0313 | 0·0390 | 0·0802 | 0·9560 |
| 35 | 0·9136 | 0·0293 | 0·0397 | 0·0174 | 0·9897 |
| | | | $R = \cdot50$ | | |
| 0 | 0·0001 | 0·0099 | 0·0099 | 0·9801 | 0·5001 |
| 10 | 0·0004 | 0·0115 | 0·0127 | 0·9755 | 0·5003 |
| 20 | 0·0007 | 0·0147 | 0·0181 | 0·9666 | 0·5006 |
| 30 | 0·0016 | 0·0208 | 0·0293 | 0·9483 | 0·5016 |
| 40 | 0·0053 | 0·0362 | 0·0610 | 0·8975 | 0·5059 |
| 50 | 0·0490 | 0·0909 | 0·2322 | 0·6279 | 0·5702 |
| 55 | 0·1944 | 0·0988 | 0·4898 | 0·2170 | 0·7986 |
| 60 | 0·3388 | 0·0335 | 0·5936 | 0·0340 | 0·9618 |
| 65 | 0·3804 | 0·0077 | 0·6054 | 0·0064 | 0·9925 |

Clearly, for both tight and loose linkage, there is a delayed sudden increase in mean fitness. For $R = 0.01$ this is associated with a sudden increase in $c_1$; for $R = 0.5$ it is associated with sudden increases in both $c_1$ and $c_3$. Note that in this particular example the sudden increase occurs earlier, but no more sharply, when linkage is tight than when it is loose. As a second example, Table 9.2 reproduces some values computed by Kimura (1965).

TABLE 9.2. Frequencies of gametes $A_1B_1$, $A_2B_1$, $A_1B_2$ and $A_2B_2$ in various generations for a model in which $w_{11} = w_{12} = w_{21} = w_{22} = 1$, $w_{13} = w_{23} = 0.99$, $w_{31} = w_{32} = 0.985$, $w_{33} = 1.02$. $R = 0.5$.

| Generation | Frequency of | | | | Change in fitness |
| | $A_1B_1$ | $A_2B_1$ | $A_1B_2$ | $A_2B_2$ | $\times 10^5$ |
| --- | --- | --- | --- | --- | --- |
| 0 | 0·2000 | 0·2000 | 0·3000 | 0·3000 | 2·926 |
| 10 | 0·2011 | 0·2040 | 0·2912 | 0·3037 | 0·661 |
| 50 | 0·1964 | 0·2239 | 0·2669 | 0·3129 | 0·455 |
| 100 | 0·1857 | 0·2426 | 0·2438 | 0·3279 | 0·370 |
| 200 | 0·1461 | 0·2594 | 0·2096 | 0·3849 | 0·933 |
| 300 | 0·0727 | 0·2290 | 0·1622 | 0·5361 | 5·71 |
| 400 | 0·0535 | 0·0846 | 0·0503 | 0·8598 | 15·29 |
| 500 | 0·0002 | 0·0641 | 0·0254 | 0·9910 | 1·68 |
| 1500 | 0·0000 | 0·0000 | 0·0000 | 1·0000 | 0·00 |

In this example there is a sudden increase in mean fitness between generations 200 and 400, after which changes in mean fitness become small. Again this sudden increase is associated with an increase in frequency of one particular gamete.

There is little point in producing further numerical examples; while it is possible to produce counter-examples, the above example together with many numerical calculations which bear on this problem show that in a large proportion of cases the 'sudden increase in mean fitness' phenomenon can be expected.

### 9.5. Linkage and the genetic load

The concept of the 'genetic load' was introduced in Section 2.4. While several types of genetic load were mentioned, we consider in this section only the segregational load. The definition (2.19) will be retained and, since in any model $w_{max}^{-1}$ is given, we confine attention to consideration of $W$, which varies more or less inversely with $L$.

If a large number of loci are kept polymorphic by what may loosely be called 'overall heterozygote advantage', the genetic load on the population might be quite considerable. It follows that it might be difficult to ascribe more than a proportion of genetic variation to this agency. Thus Lewontin and Hubby (1966) estimate about 2000 polymorphic loci in certain Drosophila populations. If each such locus is kept polymorphic because both homozygotes at each locus have fitness $0.98$, and the heterozygote at each locus has fitness unity, then the mean fitness of the population is only $0.99^{2000}$ or $10^{-9}$. Correspondingly, the genetic load $L$, defined by (2.19) is $10^9$, so that to retain stability of population size each multiple heterozygote must produce $2 \times 10^9$ viable offspring; this is clearly impossible.

This does not mean, of course, that it is impossible that at any given time there are 2000 polymorphic loci. Firstly, in large populations, a selective advantage considerably less than 2 per cent would suffice to maintain polymorphism for immense periods of time against random genetic drift. Secondly, only a proportion of the polymorphisms may be due to heterozygote advantage. The stochastic theory of previous chapters shows that a polymorphism due originally to heterozygote selective advantage may well persist for an extremely long time after such selective advantage is lost. Again, many polymorphisms are observed because the population at the time of observation is halfway through the process of replacing a less fit gene by a superior gene. Thirdly, one should keep in mind the fact that, because populations are finite, it might be extremely unlikely that a multiple heterozygote should appear. This is certainly the case with the example just quoted since the probability that any one individual is a multiple heterozygote is $(\frac{1}{2})^{2000}$ or about $10^{-600}$. This value is completely negligible and it appears reasonable to conclude that, for multiple locus arguments, a definition of genetic load which takes into account the likely composition of the population should be constructed. Thus in a population of $10^9$ individuals, the fittest individual will on average be heterozygous at only 1136 (out of 2000) loci; the implication of this is that such an individual need only reproduce about 30 offspring to maintain a stable population size.

Such considerations, however, are not our main concern at the moment. We are now primarily interested in the effect of linkage, and so far as the theory of the present chapter is concerned, this phenomenon is one which should be taken into account when discussing genetic loads at a number of loci. In particular, the above discussion has assumed that, when fitnesses are multiplicative, to a suitable approximation the mean fitness of a population is found by multiplying the mean fitnesses at the various loci. This does not take into account the possibility of equilibria (such as shown in (9.27)) at tightly linked loci, at which (8.9) does not hold. Specifically, in the model producing the equilibria (9.27), the mean population fitness at the point $c = \frac{1}{4}$ is

$$W_1 = (1 - \tfrac{1}{2}a)(1 - \tfrac{1}{2}b), \qquad (9.46)$$

(which does correspond to multiplication of single locus mean fitnesses). However, at the latter two equilibria in (9.27), the mean population fitness is

$$W_0 = W_1 - R + \tfrac{1}{4}ab,$$

which necessarily exceeds (9.46) when these equilibria exist. This implies a lower genetic load at these points. On the other hand the numerical effect is slight, and this and other numerical examples make it reasonable to conclude, when fitnesses are multiplicative, that irrespective of linkage the maintenance of a polymorphism at a large number of loci by heterozygote advantage may well entail considerable genetic load.

When fitnesses are not multiplicative, a completely different picture can emerge. Thus for the fitness matrix (9.8), single locus theory would ascribe fitnesses $1, 1 + t, 1$ to $A_1A_1, A_1A_2$ and $A_2A_2$ and hence a mean fitness of $1 + \tfrac{1}{2}t$. From a single locus point of view, the locus $B$ would be considered neutral, so that a formal multiplication of single locus fitnesses would give a mean fitness $1 + \tfrac{1}{2}t$. However, joint consideration of $A$ and $B$ loci yields at equilibrium a mean fitness $1 + \tfrac{1}{2}t + 2s(c - \tfrac{1}{4})$, where $c$ is given by (9.10), and this always exceeds $1 + \tfrac{1}{2}t$. The joint effects of linkage and epistasis thus give a lower genetic load to be carried by the population than single locus theory suggests. This result appears to be fairly general, (cf. Jain and Allard, (1966)). This appears to be particularly true of models of the 'intermediate optimum' type having a fitness matrix of the form

$$\begin{pmatrix} 1 & 1-a & 1-2a \\ 1-a & 1 & 1-a \\ 1-2a & 1-a & 1 \end{pmatrix}$$

114

For such models, equilibrium frequencies satisfy $c_1 = c_4 = \frac{1}{2} - c_2$ $= \frac{1}{2} - c_3 = c$. When linkage is tight, $c \approx \frac{1}{2}$ and $W$ is close to unity. Specifically,

$$W = 1 - 4a(\tfrac{1}{4} - c^2), \qquad (9.47)$$

where $c$ is the solution in $(0, \frac{1}{2})$ of

$$4ac^3 - 2ac^2 + R(c - \tfrac{1}{4}) = 0.$$

If $R$ is very small, this gives $c \approx \frac{1}{2} - \frac{1}{4}a^{-1}R$ and substitution in (9.47) gives $W \approx 1 - R$. Thus, over a certain range, fitness increases almost linearly as linkage is tightened. This again shows the effect of linkage on the genetic load, and again in this example it can be shown that the true genetic load is considerably less than that suggested by a composition of single locus treatments. For similar results, see Lewontin (1964b) and Nei (1965), and for a counter-example see Ewens (1968b).

Clearly, consideration of the genetic load being carried by a population must take the effect of linkage into account; further, there is some evidence that such considerations will often show that the true genetic load is rather less than suggested by a composition of single locus analyses.

## 9.6. Linkage and the survival of a new mutant

In previous chapters the survival probability of a new mutant at a locus which determines entirely the fitness of any individual was considered. In this section the analysis is extended to the case where the new mutation occurs at one of two loci which jointly determine fitness.

Specifically, we shall assume a fitness matrix of the form (8.13), and shall assume that in generation zero all individuals at $B_2 B_2$ except for one (mutant) $B_1 B_2$ individual. The problem is to find the probability of survival of the new mutant $B_1$; this is a genuine two-locus problem only if the frequencies of both $A_1$ and $A_2$ are non-negligible throughout the process.

The survival probability of the new mutant will depend on the gamete it initially forms. Since early generations are all-important, we may adopt the assumptions and notation of Section 6.2 and rewrite eqns. (9.29) and (9.30) in the form

$$E(c_1') = W^{-1}\{(W_1 - Rw_{22}q)c_1 + Rw_{22}pc_2\}. \qquad (9.48)$$

$$E(c_2') = W^{-1}\{Rw_{22}qc_1 + (W_2 - Rw_{22}p)c_2\}. \qquad (9.49)$$

Now we have seen in Section 7.3 that it is the absolute number of mutants, rather than the relative frequency of mutants, which is important for survival probabilities. The quantities $c_1$ and $c_2$ can be regarded as absolute numbers, rather than frequencies, if $W = 1$, and this equation will hold if the $w_{ij}$ are normalized suitably. In practice, this is best done by multiplying each $w_{ij}$ by $W^{-1}$, which is in effect what has been done in eqns. (9.48) and (9.49). Thus the quantities $c_1$ and $c_2$ in these equations can, and from now on will, be regarded as absolute numbers, provided that it is realized that any results deriving from these equations apply to stable populations only.

From Theorem 7.1, survival probabilities are positive only when $\lambda_1$, defined by (9.34), exceeds unity. In this case, if it is assumed that offspring distributions are Poisson, and if $\pi_i$ is the probability of survival of $B_1$, given that the initial $B_1$-carrying gamete is $A_i B_1$, then

$$-ln\ (1-\pi_1) = (W_1 - Rw_{22}q)\pi_1 + Rw_{22}q\pi_2, \qquad (9.50)$$

$$-ln\ (1-\pi_2) = Rw_{22}p\pi_1 + (W_2 - Rw_{22}p)\pi_2. \qquad (9.51)$$

There are several interesting consequences of these equations. Firstly, our knowledge of the nature of the eigenvalue (9.34) shows that there will exist cases when the survival probability of $B_1$ is positive only when $A$ and $B$ loci are sufficiently closely linked; there exist no cases where sufficiently loose linkage is necessary for survival probabilities to be positive. To this extent the stochastic theory imitates the deterministic theory. If survival probabilities are positive, eqns. (9.50) and (9.51) show that the necessary and sufficient condition that $\pi_1 = \pi_2$ is that $W_1 = W_2$. In this case survival probabilities are independent of the degree of linkage between the loci. Now the equation $W_1 = W_2$ may be written

$$p = (w_{23} - w_{22})/(w_{21} + w_{23} - 2w_{22}), \qquad (9.52)$$

and if $A_1$ and $A_2$ had been kept at non-negligible frequencies through heterozygote advantage of $A_1 A_2$, the frequency $p$ of $A_1$ would be

$$p = (w_{33} - w_{32})/(w_{31} + w_{33} - 2w_{32}). \qquad (9.53)$$

Equations (9.52) and (9.53) will hold simultaneously if and only if constants $a$ and $b$ can be found such that

$$w_{2i} = a + bw_{3i}, \qquad (i = 1, 2, 3). \qquad (9.54)$$

Particular cases of fitnesses of the form (9.54) are

$$w_{ij} = u_i v_j, \qquad w_{ij} = u_i + v_j. \tag{9.55}$$

Thus if either of these representations of $w_{ij}$ is possible, and if the original polymorphism $A_1 - A_2$ was maintained by heterozygote superiority, then the survival probability of $B_1$ is independent of the degree of linkage between $A$ and $B$ loci, and also of the initial $B_1$-carrying gamete. To this extent, and despite the conclusions of Chapter 8, either representation (9.55) can be regarded as defining independence of $A$ and $B$ loci.

Since the initial $B_1$-carrying gamete will be $A_1 B_1$ with probability $p$ and $A_2 B_2$ with probability $q$, it is pertinent to examine the nature of the weighted survival probability $\pi = p\pi_1 + q\pi_2$; this is best done by examining separately the three cases (i) $W_1 > W$, $W_2 > W$, (ii) $W_1 > W$, $W_2 < W$, $W^* < W$, (iii) $W_1 > W$, $W_2 < W$, $W^* > W$.

In case (i) both $\pi_1$ and $\pi_2$ are positive when $R = 0$; writing the solutions of (9.50) and (9.51) in this case as $\pi_1^*$ and $\pi_2^*$, it follows that

$$1 - \pi_i^* = \exp\{-(W_i/W)\pi_i^*\}, \qquad (i = 1, 2). \tag{9.56}$$

Equations (9.50), (9.51) and (9.56) now show jointly that

$$\left[\frac{d\pi}{dR}\right]_{R=0} = pqw_{22}W^{-1}(\pi_2^* - \pi_1^*)\left[\frac{1 - \pi_1^*}{(1 - (1 - \pi_1^*)W_1 W^{-1}}\right.$$
$$\left. - \frac{1 - \pi_2^*}{1 - (1 - \pi_2^*)W_2 W^{-1}}\right]. \tag{9.57}$$

Using eqn. (9.56) again, both sides in this equation will be positive if

$$\frac{1 - \pi_1^*}{1 + (\pi_1^*)^{-1}(1 - \pi_1^*)\log(1 - \pi_1^*)} > \frac{1 - \pi_2^*}{1 + (\pi_2^*)^{-1}(1 - \pi_2^*)\log(1 - \pi_2^*)} \tag{9.58}$$

whenever $\pi_2^* > \pi_1^*$. But this is always the case since

$$(1 - x)[1 + x^{-1}(1 - x)\log(1 - x)]^{-1}$$

is a decreasing function of $x$ for $x$ in $(0, 1)$. The quite unexpected conclusion is thus reached that despite the result of Section 2, the probability of survival of a new mutant is never maximized in case (i) by extremely close linkage.

The mathematical reason for this apparently contradictory

behaviour is obvious enough. The deterministic theory depends solely on the value of the eigenvalue (9.34), whereas the stochastic theory depends on linear combinations of powers of both eigenvalues (9.34) and (9.35). Although both eigenvalues decrease with $R$, it so happens that the linear combination generally throws more weight on the larger eigenvalue as $R$ increases. Often there appears to be a moderate value for $R$ where an optimum combination of decreasing eigenvalues and increasing weight on the larger eigenvalue is reached; such a case will be illustrated in a numerical example below.

Genetically, the consequence of this behaviour is that in any problem concerning rare mutants, it must be known whether the mutant is so rare that a stochastic treatment is necessary, since this may differ somewhat from a deterministic treatment.

A numerical example illustrating this behaviour is provided when the fitness matrix (8.13) takes the form

$$\begin{pmatrix} * & * & * \\ 1\cdot 0 & 1\cdot 2 & 1\cdot 3 \\ 1\cdot 0 & 1\cdot 1 & 1\cdot 0 \end{pmatrix} \tag{9.59}$$

with $p = q = \frac{1}{2}$. For these numerical values, $W = 1\cdot 05$, $W_1 = 1\cdot 1$, $W_2 = 1\cdot 25$ and $W^* = 1\cdot 175$. Values of $\pi_1$, $\pi_2$ and $\pi$ for various $R$, together with values of $E(c_1^{(+)}) + E(c_2^{(t)})$, are shown below in Table 9.3.

A second point, not so directly concerned with the effect of linkage, is to examine how good an approximation is provided by a formal

TABLE 9.3. Survival probabilities for the model (9.59).

| $R$ | $\pi_1$ | $\pi_2$ | $\pi$ |
|---|---|---|---|
| $\approx 0$ | 0·090 | 0·302 | 0·196 |
| 0·1 | 0·158 | 0·271 | 0·215 |
| 0·5 | 0·192 | 0·233 | 0·213 |

| | Expected number of gametes in generation $t$ | |
|---|---|---|
| $R$ | $A_1B_1$ | $A_2B_1$ |
| $\approx 0$ | $0(1\cdot 1905)^t + 1(1\cdot 0476)^t$ | $1(1\cdot 1905)^t + 0(1\cdot 0476)^t$ |
| 0·1 | $0\cdot 4219(1\cdot 1534)^t + 0\cdot 5781(0\cdot 9704)^t$ | $1\cdot 2028(1\cdot 1534)^t - 0\cdot 2028(0\cdot 9704)^t$ |
| 0·5 | $0\cdot 8639(1\cdot 1278)^t + 0\cdot 1361(0\cdot 5388)^t$ | $1\cdot 1063(1\cdot 1278)^t - 0\cdot 1063(0\cdot 5388)^t$ |

single locus treatment. Single locus theory would ascribe fitnesses
1·05 and 1·175 to $B_2B_2$ and $B_1B_2$, and thus would ascribe a survival
probability $\pi$ for $B_1$ satisfying the equation

$$-ln\,(1-\pi) = 1·175\pi/1·05.$$

The solution of this equation is $\pi = 0·205$, which falls within the span
of the values found under two-locus theory by varying $R$. Other
numerical examples indicate that this is a fairly general result,
although direct mathematical confirmation of this seems rather
difficult.

In cases (ii) and (iii), $\pi_2 = 0$ whenever $R = 0$, so that

$$\left[\frac{d\pi}{dR}\right]_{R=0} = pqw_{22}W^{-1}\pi_1{}^*\left[\frac{W}{W-W_2} - \frac{1-\pi_1{}^*}{1-(1-\pi_1{}^*)W_1W^{-1}}\right].$$

After some algebra, it is found that the condition that the right-hand
side be positive is

$$W^* > W + W[\tfrac{1}{2}\pi_1{}^* + \tfrac{1}{3}(\pi_1{}^*)^2 + \ldots - q\{\pi_1{}^* + (\pi_1{}^*)^2 + \ldots\}].$$

$$(9.60)$$

It is clear that (9.60) can hold even if $W^* < W$, and conversely that
(9.60) need not hold even when $W^* > W$. That is to say, it is not always
true that in case (ii) optimal values for survival probabilities are provided by extremely tight linkage, nor is it true that in case (iii) such
optimal values cannot occur for $R = 0$. To exemplify this, suppose
that the fitness matrix (8.13) is

$$\begin{pmatrix} * & * & * \\ 1·4 & 1·0 & 0·9 \\ 0·9775 & 1·0075 & 0·9975 \end{pmatrix}$$

with $p = \tfrac{1}{4}, q = \tfrac{3}{4}$ (balanced polymorphism for $A_1 - A_2$ in the absence
of $B_1$). With these values $W = 1$, $W^* = 0·9688$ (so that case (ii)
applies). For $R = 0\ \pi = 0·044034$ while for $R = 0·04$, $\pi = 0·04411$.
Thus, remarkably, while survival probabilities can be positive only
when $R < 0·24$, the optimal value of $R$ for survival probabilities is not
$R = 0$.

Continuing this line of thought, it is clear that whenever $W$ and
$W^*$ are close, the requirement (9.36) will be of the form $R < K$, where
$K$ might be quite large. But only values of $R$ less than $\tfrac{1}{2}$ have any
genetical significance so that it is possible even when $W^* < W$, the

weighted survival probability for $R = \frac{1}{2}$ exceeds that for $R = 0$. Such behaviour occurs when (8.13) is

$$\begin{pmatrix} * & * & * \\ 1\cdot5 & 1\cdot5 & 0\cdot9873 \\ 0\cdot901 & 1\cdot001 & 1\cdot000 \end{pmatrix}$$

with $p = 0\cdot01$, $q = 0\cdot99$. Here $W = 1$, $W^* = 0\cdot9975$ and (9.36) becomes $R < 1$. For $R = 0$, $\pi = 0\cdot0058$ while for $R = \frac{1}{2}$, $\pi = 0\cdot0126$. For examples such as this, it is quite clear that a deterministic analysis gives a quite misleading idea of the effect of linkage on the survival probabilities of new mutants.

Finally, when (8.13) is

$$\begin{pmatrix} * & * & * \\ 1\cdot1333 & 1\cdot0 & 0\cdot4 \\ 0\cdot9975 & 1\cdot0075 & 0\cdot9775 \end{pmatrix}$$

with $p = \frac{3}{4}$, $q = \frac{1}{4}$, then $W = 1$ and $W^* = 1\cdot0375$ so that case (iii) applies. However the requirement (9.60) does not hold, so that for example $\pi = 0\cdot1321$ when $R = 0$ and yet $\pi = 0\cdot1285$ when $R = 0\cdot04$.

Clearly the effect of linkage on survival probabilities can be a little unexpected. Close linkage does not necessarily provide the maximum value; indeed it never does when both $W_1$ and $W_2$ exceed $W$. If (say) $W_2 < W < W_1$, close linkage is optimal only when (9.60) does not hold; otherwise moderate linkage is optimal. There is no particular relationship between the value of $W^*$ and whether or not close linkage is optimal.

# CHAPTER 10

# Dominance

The recognition of the phenomenon of dominance is as old as the Mendelian theory itself; there still does not appear to be, however, anything approaching general agreement on the reason for its existence (cf. Crosby (1963), Clark (1964), Ford and Sheppard (1965), Wright (1964)). In this chapter an account is given of the theory of Fisher, the discussion of which is largely quantitative, that dominance appears as a result of natural selection. In this context, to say that $A_1$ is dominant to $A_2$ is to say that the heterozygote $A_1A_2$ is indistinguishable in all respects from the homozygote $A_1A_1$; in particular it follows that the heterozygote has the same fitness as this homozygote.

## 10.1. Dominance of the wild type

In nature it is frequently found that a particular gene occurs with extremely high frequency in a population; indeed the appearance of other genes seems to be due only to recurrent mutation from this 'wild type' gene. Very often it happens that such a wild type is dominant to the genes which arise from it by mutation, and it was with this particular fact in mind that Fisher (1928a, b) put forward a theory of the evolution of dominance by natural selection of modifiers, which may in brief be described as follows.

If $A_1$ is the wild type and $A_2$ a rare mutant, then heterozygotes $A_1A_2$, although rare, will be far more common than homozygotes $A_2A_2$. Suppose now there exists a second locus '$M$', admitting alleles $M_1$ and $M_2$, whose effect is such that the fitness (and phenotypic expression) of those $A_1A_2$ individuals carrying the modifying gene $M_1$ is altered towards that of $A_1A_1$ (who are not themselves modified in any way). Thus $A_1A_2M_1$-individuals are more fit than $A_1A_2M_2M_2$ individuals; this means that $M_1$ is at an (induced) selective advantage to $M_2$ and will therefore steadily increase in frequency (provided that no other selective forces act). Eventually $M_1$ will become fixed in the population, and since this will imply that all $A_1A_2$ individuals are indistinguishable from $A_1A_1$, dominance of $A_1$ over $A_2$ in the population will have been achieved.

We consider now a quantitative treatment of this process. Suppose that in the absence of the modifier $M_1$, the fitnesses of $A_1A_1$, $A_1A_2$ and $A_2A_2$ are 1, $1-sh$, $1-s$ respectively, where $s>0$ and $0 \leq h \leq 1$. If the mutation rate $A_1 \rightarrow A_2$ is $u$, then to a sufficient approximation

$$x = \text{frequency } (A_1) = 1-(u/sh). \tag{10.1}$$

Suppose now that the fitness matrix (8.13) is

$$\begin{pmatrix} & A_1A_1 & A_1A_2 & A_2A_2 \\ M_1M_1 & 1 & 1 & 1-s \\ M_1M_2 & 1 & 1-sk & 1-s \\ M_2M_2 & 1 & 1-sh & 1-s \end{pmatrix} \tag{10.2}$$

Here it is assumed that $0 \leq k \leq h$; if $k = 0$ the modifier $M_1$ acts as a dominant in its modifying action. It is worth while noting some qualitative facts about processes governed by a fitness matrix such as (10.2). Firstly, if $s$ is very small, nearly all fitnesses are identical and only very slow changes in gene frequency will occur. If linkage between $A$ and $M$ loci is not tight, the theory of the previous chapter suggests that to a very close approximation the frequency of any gamete can be taken as the product of the frequencies of the constituent genes. This approximation will be made for the moment; later a more complete analysis will discuss its adequacy.

Secondly, while interest centres mainly on changes in the frequency of $M_1$, the frequency of $A_1$ will also change as the process continues. Thus as the frequency of $M_1$ increases from zero to one, the frequency $x$ of $A_1$ will steadily change from that given by (10.1) to the mutation-selection point applicable when the frequency of $M_1$ is unity, namely $1-\sqrt{(u/s)}$. Suppose that at any arbitrary time the frequency of $M_1$ is $y$. Then, from (8.15)–(8.18),

$$\Delta y = sx(1-x)y(1-y)\{4ky+2h-2hy-2k\}, \tag{10.3}$$

$$\Delta x = sx(1-x)\{1-x+(2x-1)(1-y)(2ky+h-hy)\}-ux. \tag{10.4}$$

The final term in (10.4) does not come from (8.15)–(8.18) but arises from the recurrent mutation $A_1 \rightarrow A_2$.

The main problem is to find an expression for $\Delta y$ which is independent of $x$, since such an expression gives an indication of the selecitve force acting on the modifier $M_1$ through the process of dominance modification. As a reasonable first approximation, eqns. (10.3) and (10.4) together form the differential equation

$$\frac{dx}{dy} = \frac{sx(1-x)\{1-x+(2x-1)\,(1-y)\,(2ky+h-hy)\}-ux}{sx(1-x)y(1-y)\{4ky+2h-2hy-2k\}}. \quad (10.5)$$

If this could be solved for $x$ as a function of $y$, the expression for $x$ so obtained could be inserted in (10.3), thus providing an expression for $\Delta y$ in terms of $y$ only. Unfortunately, this approach breaks down since it appears to be impossible to solve eqn. (10.5), and progress in this direction can only be made by solving (10.5) numerically for various representative cases.

A slightly less accurate method is to argue as follows. While it is not true that $\Delta x = 0$, the value of $x$, for any given value of $y$, should be close to that value which is in selection-mutation equilibrium for the chosen value of $y$. Such a value is found by putting $\Delta x = 0$, or by putting

$$(1-x)^2+(1-x)\,(2x-1)\,(1-y)\,(2ky+h-hy) = u/s. \quad (10.6)$$

Since $x$ is always very close to unity, one is tempted to ignore the first term in (10.6) and, by replacing $2x-1$ by 1, to find a value for $1-x$ which can be used in (10.3). Unfortunately, when $y$ is near 1, the first term dominates the second term and this approach breaks down. However, very little accuracy is lost by replacing the term $2x-1$ in (10.6) by 1, since the error involved in doing so is of order $(1-x)^2(1-y)$ and is therefore always extremely small. To this approximation the solution of (10.6) is

$$1-x = \tfrac{1}{2}[-(1-y)\,(2ky+h-hy)+\{(1-y)^2(2ky+h-hy)^2+4u/s\}^{\frac{1}{2}}].$$

This value, together with the approximation $x \approx 1$, when substituted in (10.3) yields

$$\Delta y = \tfrac{1}{2}sy(1-y)\,[4ky+2h-2hy-2k]\,[-(1-y)\,(2ky+h-hy)$$
$$+\{(1-y)^2\,(2ky+h-hy)^2+4u/s\}^{\frac{1}{2}}]. \quad (10.8)$$

In the important particular case $k = 0$ this becomes

$$\Delta y = shy(1-y)\,[-h(1-y)^3+(1-y)\{h^2(1-y)^4+4u/s\}^{\frac{1}{2}}], \quad (10.9)$$

and when $k = \tfrac{1}{2}h$,

$$\Delta y = \tfrac{1}{2}shy(1-y)\,[-h(1-y)+\{h^2(1-y)^2+4u/s\}^{\frac{1}{2}}]. \tag{10.10}$$

Clearly, in both cases, $\Delta y$ increases from zero at $y = 0$, reaches a maximum for some value of $y$ in $(0,1)$, and then decreases to zero.

Suppose now that the modifier $M_1$ is subject to selective influences other than those derived from its dominance modification action. In the simplest case these forces produce a change

$$\Delta y = Cy(1-y) \tag{10.11}$$

in the frequency of the modifier. When (10.11) is true, the constant $C$ may be called the 'ordinary' selective force acting on the modifier, and it is therefore important to find the value corresponding to $C$ derived from dominance modification. Equations (10.9) and (10.10) show that the value of this force depends on $y$, so that it is reasonable to consider the maximum of this force for various $y$. When $k = 0$, this maximum occurs where

$$1-y = h^{-\frac{2}{3}}(4u/3s)^{\frac{1}{3}}, \tag{10.12}$$

and at this point assumes the value

$$1 \cdot 24 h^{\frac{1}{3}} u^{\frac{2}{3}} s^{\frac{1}{3}}. \tag{10.13}$$

This implies that, taking the forces due to dominance modification only into account,

$$\Delta y \leqq y(1-y)\{1 \cdot 24 h^{\frac{1}{3}} u^{\frac{2}{3}} s^{\frac{1}{3}}\}. \tag{10.14}$$

When $k = \tfrac{1}{2}h$, similar arguments yield

$$\Delta y \leqq y(1-y)\{h(us)^{\frac{1}{2}}\}. \tag{10.15}$$

Before drawing any conclusions from these results, the large number of approximations used means that some check should be made on their accuracy. Such a check would involve the numerical computation of gamete frequencies using the recurrence relations (8.15)–(8.18). The only calculations available for this purpose refer to the case $k = 0$, $h = \tfrac{1}{2}$, $s = 0\cdot2$, $u = 2 \times 10^{-5}$ and have been computed by Kimura (and reproduced by Ewens, (1967)). The values, together with those computed from (10.8) and the numerical solution of (10.5), are reproduced in Table 10.1.

It is clear from this table that the numerical values found from eqn. (10.9), while not completely accurate, nevertheless provide an extremely close approximation for the selective force acting on the

TABLE 10·1. Values of the selective force ($\times 10^5$) acting on a
modifier through dominance modification.

$k=0$, $h=\frac{1}{2}$, $s=0.2$, $u=2\times10^{-5}$.

| Frequency of modifier | Exact value of selective force | Value found from (10.8) | Value found from (10.5) |
|---|---|---|---|
| 0·1 | 3·599 | 4·442 | 3·600 |
| 0·2 | 4·996 | 4·995 | 4·996 |
| 0·3 | 5·704 | 5·705 | 5·706 |
| 0·4 | 6·654 | 6·646 | 6·650 |
| 0·5 | 7·948 | 7·950 | 7·952 |
| 0·6 | 9·802 | 9·848 | 9·854 |
| 0·7 | 12·54 | 12·73 | 12·62 |
| 0·8 | 15·97 | 16·56 | 16·14 |
| 0·9 | 14·99 | 15·62 | 15·20 |
| 0·95 | 9·317 | 9·393 | 8·652 |
| 0·98 | 3·896 | 3·600 | 3·912 |

modifier. Equation (10.12) suggests that the selective force reaches a
maximum when $y = 0·848$, and at this point has the value $17·54 \times 10^{-5}$; these values agree very well with the numerical results. It is
therefore reasonable to use (10.12)-(10.15) to discuss the effect of
dominance modification on the modifying gene $M_1$.

The most obvious observation to make is that the selective effect of
this modification is always extremely small. Normally, values of $u$
are of order $10^{-5}$ or $10^{-6}$ while values of $s$, except in extreme cases,
are 0·1 or less. Thus when $k = 0$ the selective force acting on the
modifier through dominance modification can seldom exceed 0·0001,
and for most values of $y$ must be rather less than this. When $k = \frac{1}{2}$ the
selective force can seldom exceed 0·001, and again will usually be far
less. It is here that the mathematical treatment must stop, since the
question to be answered is now whether, if dominance is caused by
several modifiers, there exist in nature modifying genes which are
otherwise so neutral that a selective force of order 0·001 or 0·0001 will
be the major one on the gene; this is clearly not a mathematical
question.

These considerations raise the question of what happens when there
exists a number of modifying loci. The algebraic problems here are
now quite difficult and resort must be made to numerical computation.

Some results of Mayo (1966) suggest that a number of minor modifiers acting together are less effective than an 'equivalent' major modifier; if this is so, the numerical values found about could be used as upper bounds. Further results have been obtained by O'Donald (1967a, b) using a quite independent approach. Again, whatever further models suggest, the eventual answers cannot be provided by a mathematical treatment alone.

## 10.2. Primary genes which are initially rare

The reason for the comparative smallness of the selective force due to dominance modification is the comparative rarity of heterozygotes $A_1A_2$. This raises the question of what happens if $A_1$ is initially rare and then for some reason, possibly a changed environment, steadily increases in frequency to unity. During the course of this increase there will be a large number of heterozygotes upon which dominance modification can act, bringing about a possible substantial increase in the frequency of the modifier.

Questions of this sort must be answered numerically rather than algebraically, since it is now not sufficiently accurate to consider gene frequencies only. Table 10.2 presents the progress of four populations where gamete frequencies are determined by models of the form (10.2), together with the recurrence relations (8.15)–(8.18).

While in all cases the frequency of the modifier increases, the increase is larger when linkage between primary and modifying loci is tight and also when selective differences at the primary locus are large. These are general results and could have been anticipated from the previous chapter. It does appear reasonable to state, however, that unless linkage is very tight and unless selective differences are very large, the phenomenon under consideration can have little effect in raising significantly the frequency of the modifier. Even if the frequency of the modifier were increased significantly, the considerations of the previous section would become relevant and the selective force acting on the modifier will usually be majorized by other selective pressures.

## 10.3. Dominance and the genetic load

In this section we consider an entirely novel and extremely interesting theory advanced by Kimura (1960) which could account for the occurrence of dominance (and of other phenomena). Briefly, Kimura

TABLE 10.2. Values of gamate and gene frequencies for various $h$, $k$, $s$ (see (10.2)) and various linkage values.

$$s = 0{\cdot}1, h = 1, k = 0{\cdot}5$$

| Generation | $A_1M_1$ | $A_2M_1$ | $A_1M_2$ | $A_2M_2$ | $M_1$ | $A_1$ |
|---|---|---|---|---|---|---|
| | | | $R = 0{\cdot}02$ | | | |
| 0 | 0·0001 | 0·0099 | 0·0099 | 0·9801 | 0·0100 | 0·0100 |
| 50 | 0·0009 | 0·0105 | 0·0111 | 0·9776 | 0·0113 | 0·0119 |
| 100 | 0·0056 | 0·0132 | 0·0147 | 0·9665 | 0·0188 | 0·0203 |
| 150 | 0·0347 | 0·0291 | 0·0332 | 0·9029 | 0·0639 | 0·0680 |
| 200 | 0·2267 | 0·1016 | 0·1344 | 0·5374 | 0·3283 | 0·3611 |
| 250 | 0·6172 | 0·0960 | 0·2449 | 0·0148 | 0·7132 | 0·8622 |
| 295 | 0·7217 | 0·0423 | 0·2316 | 0·0043 | 0·7640 | 0·9534 |
| | | | $R = 0{\cdot}5$ | | | |
| 0 | 0·0001 | 0·0099 | 0·0099 | 0·9801 | 0·0100 | 0·0100 |
| 100 | 0·0002 | 0·0112 | 0·0128 | 0·9758 | 0·0114 | 0·0128 |
| 200 | 0·0003 | 0·0132 | 0·0170 | 0·9696 | 0·0135 | 0·0172 |
| 300 | 0·0005 | 0·0170 | 0·0261 | 0·9563 | 0·0175 | 0·0267 |
| 400 | 0·0012 | 0·0234 | 0·0438 | 0·9316 | 0·0247 | 0·0450 |
| 500 | 0·0073 | 0·0454 | 0·1225 | 0·8248 | 0·0527 | 0·1298 |
| 550 | 0·0487 | 0·0758 | 0·3369 | 0·5385 | 0·1246 | 0·3856 |
| 600 | 0·2507 | 0·0184 | 0·6849 | 0·0459 | 0·2892 | 0·9356 |

$$s = 0{\cdot}5, h = 1, k = 0{\cdot}5$$

| Generation | $A_1M_1$ | $A_2M_1$ | $A_1M_2$ | $A_2M_2$ | $M_1$ | $A_1$ |
|---|---|---|---|---|---|---|
| | | | $R = 0{\cdot}02$ | | | |
| 0 | 0·0001 | 0·0099 | 0·0099 | 0·9801 | 0·0100 | 0·0100 |
| 5 | 0·0007 | 0·0102 | 0·0107 | 0·9784 | 0·0109 | 0·0114 |
| 10 | 0·0053 | 0·0107 | 0·0118 | 0·9721 | 0·0160 | 0·0172 |
| 15 | 0·0366 | 0·0129 | 0·0148 | 0·9356 | 0·0495 | 0·0515 |
| 20 | 0·2037 | 0·0231 | 0·0266 | 0·7466 | 0·2269 | 0·2303 |
| 25 | 0·5790 | 0·0426 | 0·0503 | 0·3280 | 0·6217 | 0·6294 |
| 30 | 0·8122 | 0·0478 | 0·0622 | 0·0778 | 0·8600 | 0·8744 |
| 35 | 0·8796 | 0·0425 | 0·0628 | 0·0150 | 0·9222 | 0·9425 |
| 40 | 0·9000 | 0·0363 | 0·0609 | 0·0029 | 0·9363 | 0·9609 |
| | | | $R = 0{\cdot}5$ | | | |
| 0 | 0·0001 | 0·0099 | 0·0099 | 0·9801 | 0·0100 | 0·0100 |
| 10 | 0·0004 | 0·0115 | 0·0127 | 0·9755 | 0·0118 | 0·0130 |
| 20 | 0·0007 | 0·0147 | 0·0181 | 0·9666 | 0·0154 | 0·0188 |
| 30 | 0·0016 | 0·0208 | 0·0293 | 0·9483 | 0·0224 | 0·0308 |
| 40 | 0·0053 | 0·0362 | 0·0610 | 0·8975 | 0·0415 | 0·0663 |
| 50 | 0·0490 | 0·0909 | 0·2322 | 0·6279 | 0·1399 | 0·2812 |
| 60 | 0·3388 | 0·0335 | 0·5936 | 0·0340 | 0·3724 | 0·9325 |

argues that phenomena observed today occur because they are in some sense 'optimum'; more specifically, they are 'optimum' because they lead to a minimization of the overall genetic load.

So far in this book attention has been concentrated mainly on the segregational genetic load, the mutational genetic load being mentioned only briefly in Section 2.4. Suppose that the genotypes $A_1A_1$, $A_1A_2$ and $A_2A_2$ have fitnesses 1, $1-sh$, $1-s$, where $0 < sh < s$. Optimum fitness is achieved when all individuals are $A_1A_1$, but the effect of recurrent mutation (at rate $u$) from $A_1$ to $A_2$ is that the equilibrium frequency of $A_1$ is slightly less than unity and a small mutational load is incurred. It was mentioned in Chapter 2 that this load is approximately $2u$; more precisely, the equilibrium frequency $x_0$ of $A_2$ satisfies (for small $s$) the equation

$$u = sx_0\{h + (1 - 2h)x_0\}, \tag{10.16}$$

and to the same order of accuracy, the mutational load is

$$sx_0\{x_0 + 2h(1 - x_0)\}. \tag{10.17}$$

A third type of genetic load, not yet discussed, is the 'evolutionary load'. Suppose that, because of environmental changes, $s$ were to become negative; the frequency of $A_2$ would then steadily increase. If at any stage during this increase the frequency of $A_2$ is $x$, then to the order of accuracy considered the genetic load (where now $w_{\max} = 1 + |s|$) is

$$-s(1 - x)(1 + x - 2hx). \tag{10.18}$$

Thus as $x$ increases from $x_0$ to unity, a total 'evolutionary' genetic load may be defined as

$$-\int_0^\infty s(1 - x)(1 + x - 2hx)dt, \tag{10.19}$$

where $t$ is time measured in generations. Since, approximately,

$$dx/dt = -sx(1 - x)\{h + (1 - 2h)x\},$$

this expression may be replaced by

$$\int_{x_0}^1 [1 + x - 2hx][x\{h + (1 - 2h)x\}]^{-1}dx$$
$$= -h^{-1}[\log x_0 + (1 - h)\log\{(1 - h)/(h + (1 - 2h)x_0)\}]. \tag{10.20}$$

128

The total genetic load is not simply the sum of (10.17) and (10.20). These two components apply at different times, and while (10.17) might apply for very long periods while the environment remains constant, (10.20) will apply for what might be the comparatively short time during which marked changes in gene frequencies occur. It is reasonable then to multiply (10.20) by some constant $c$ before adding to (10.17) to get a total genetic load, where $c$ is the mean rate at which changes in environment induce, for any given locus, a change in gene frequency.

These considerations lead to the following mathematical problem: subject to the constraint (10.16), minimize the expression

$$sx_0\{x_0+2h(1-x_0)\}-ch^{-1}[\log x_0+(1-h)\log\{(1-h)/(h+(1-2h)x_0)\}]$$
(10.21)

with respect to variation in $s$, $h$ and $u$. This has been done by Kimura who finds that at the minimizing values the following equations, in addition to (10.16), hold:

$$2h\{h+x_0(1-2h)\}-(c/s)[x_0^{-1}-(1-h)(1-2h)/\{h+(1-2h)x_0\}]=0,$$
(10.22)

$$2hx_0(1-x_0)+(c/s)[1+h^{-1}\log\{(1-hx_0/(h+x_0-2hx_0)\}]$$
$$+(1-h)(1-2x_0)/\{h+x_0-2hx_0\}=0.$$
(10.23)

These equations must be solved numerically; any given value of either of the parameters $h$, $x_0$ or $c/s$ will automatically fix the remaining two. As an indication of the sort of results obtained, some solutions of eqns. (10.16), (10.22) and (10.23) for typical values are given in Table 10.4.

### TABLE 10.4

| $x_0$ | $1\times10^{-3}$ | $5\times10^{-3}$ | $1\times10^{-2}$ | $2\times10^{-2}$ | $5\times10^{-2}$ | $1\times10^{-1}$ |
|---|---|---|---|---|---|---|
| $h$ | $2{\cdot}172\times10^{-3}$ | $1{\cdot}104\times10^{-2}$ | $2{\cdot}252\times10^{-2}$ | $4{\cdot}664\times10^{-2}$ | $1{\cdot}259\times10^{-1}$ | $2{\cdot}640\times10^{-1}$ |
| $c/s$ | $2{\cdot}004\times10^{-8}$ | $2{\cdot}525\times10^{-6}$ | $2{\cdot}037\times10^{-5}$ | $1{\cdot}684\times10^{-4}$ | $2{\cdot}570\times10^{-3}$ | $1{\cdot}849\times10^{-2}$ |

The values in this table suggest that $x_0$ is approximately $\frac{1}{2}h$; a power series expansion for $x_0$ (in terms of $h$) reveals that a more accurate approximation is

$$x_0\approx0{\cdot}4624h-0{\cdot}914h^2+5{\cdot}232h^3-19{\cdot}14h^4+\ldots\ldots$$
(10.24)

This power series expansion, when used in (10.22), (10.16) and (10.17) gives

$$c/s \approx 1 \cdot 978h^3 - 4 \cdot 073h^4 + \ldots \ldots , \tag{10.25}$$

$$u \approx 0 \cdot 342ch^{-1}\{1 - 0 \cdot 895h + \ldots \ldots\}, \tag{10.26}$$

$$sx_0 \approx uh^{-1}\{0 \cdot 684 + 0 \cdot 860h + \ldots \ldots\}. \tag{10.27}$$

Summing (10.26) and (10.27) over all loci,

$$\sum u \approx 0 \cdot 342Ch^{-1}\{1 - 0 \cdot 895h + \ldots \ldots\}, \tag{10.28}$$

$$\sum (sx_0) \approx (\sum u)h^{-1}\{0 \cdot 684 + 0 \cdot 860h + \ldots \ldots\}, \tag{10.29}$$

where $C = \sum c$ is the mean total number of loci for which, in any generation, a process of gene substitution begins. The value of (10.28) and (10.29) is that independent estimates exist for $\sum(sx_0)$ and $C$. The former is called the 'total mutational damage per gamete' and has been estimated by Morton, Crow and Muller (1956) to be approximately 2 in human populations while the quantity $C$ has been estimated by Haldane (1957) at about 1/300. Use of these two estimates in (10.28) and (10.29) yields

$$h \approx 0 \cdot 02, \qquad (\textstyle\sum u) \approx 0 \cdot 06,$$

and further argument yields an average value for $s$ at between 1 per cent and 2 per cent. These estimates for $(\sum u)$ and $s$ agree well with observation, but for the moment our main interest is in the mean value predicted for $h$. The result that the optimum value for $h$ is quite small implies that by natural selection, populations having small values of $h$ will tend to survive; that is to say, natural selection favours survival of populations exhibiting almost complete dominance. While there are several reservations one may wish to make about various assumptions in the above analysis, it is nevertheless extremely suggestive and interesting and provides a theory of the evolution of dominance independent of, and alternative to, that of Fisher.

It should be mentioned in conclusion that the problem of dominance illustrates a situation which has occurred several times in population genetics. This is that while its original consideration was quantitative, the various theories (based on quantitative arguments) have led to a considerable amount of fieldwork being undertaken, the results of which now appear to be more important than the original quantitative analysis.

# CHAPTER 11

# Summary

In this final chapter, an attempt will be made to bring together the more significant conclusions reached previously and to try to see whether these lead to any general principles of evolution, and if so, what these principles are.

In the first place, it must be understood that, because the biological world is infinitely more complex than our mathematical models, it is impossible to expect that mathematics can play in the biological sciences the fundamental and ubiquitous role which it plays in the physical sciences. We must therefore expect that such principles as do emerge must be treated with some measure of caution and with full appreciation of the difficulty of forming workable and meaningful mathematical models in the biological world. On the other hand, one should not forget the possibility mentioned at the end of the previous chapter, that mathematical investigation which subsequently appears to be of lesser importance initiates experiments and fieldwork from which important results flow. This has happened sufficiently often to be always kept in mind as an important by-product of a mathematical approach.

Despite the problems mentioned above, it seems impossible to deny that a number of the results we have obtained are of considerable importance. Firstly, irrespective of any possible unreality of the assumptions under which the Hardy-Weinberg law was derived, the general importance of this law is undeniable. The Mendelian system of heredity ensures the existence of a vast and continuing store of variability upon which the forces of natural selection can act. The general conclusion of the stochastic treatment of Chapters 4, 5 and 6 is that a population must be fairly small before such a conclusion can no longer be drawn. When no selective forces act, populations of size about $10^6$ are sufficiently large for random effects to be ignored; if selective forces act the number will be considerably reduced. Thus if selective differences are of the order $0.1$, populations having only several hundred individuals can be considered to be 'large'.

Secondly, while the assumptions under which the 'Fundamental Theorem of Natural Selection' were obtained are again no doubt idealistic, and despite the two-locus theory results in Chapters 8 and 9, it is probably true that in the great majority of cases, the action of selection is such that the mean fitness of the population increases from one generation to the next and that as a result the population is in some sense 'improved'. Further, the rate of such 'improvement' will usually be proportional to some component of the genetic variance. Again some qualification must be made in the case of small populations, for which the effect of random sampling should be taken into account. When this is done, the results of Section 6.1 become important. The main conclusion of this section, (typified by eqn. (6.1)), is that even a very small selective difference can play a decisive effect in the evolutionary behaviour of a population. Indeed the mathematical results show that the effect of selection is far greater than intuitive ideas might initially suggest. Taken in conjunction with the calculations in Chapters 2 and 6 concerning the times required for given changes in gene frequency, one might describe the effect of selection in Mendelian systems as being often slow but nearly always sure, which, all things considered, is probably the most desirable way of operating.

Perhaps the most striking effects of selection occur when linkage is taken into account. Even simple models such as (9.15) show how complicated the joint effects of linkage and selection can be. If any general conclusion is to be drawn from such models, it would be that when linkage is loose, the behaviour of the population will often approximate that suggested by a composition of single locus analyses, but that when linkage is tight this is no longer the case, the behaviour being far less simple. Since the linkage systems which do occur in nature will be far more complex than those considered here, the possible complications of the effect of selection in such systems can hardly be imagined. Indeed, of all the factors which might lead to one to be dubious about the relevance of the single locus theory comprising a large portion of the theory of population genetics, this is by far the most outstanding. On the other hand, it appears that certain problems associated with the survival probabilities of new mutants do not lose relevance when treated as single locus rather than multi-locus problems. Again, the single locus treatment of the problem of self-sterility alleles would probably change little under a multi-locus approach.

132

From the genetical point of view, there are also many cases where selective effects depend to a large extent on a very small number of loci; the case where the theory outlined above will most likely fail is where the total force of selection depends on the composition of a large number of rather small selective differences at a large number of loci.

The three men who initiated and whose views have dominated the study of the field of population genetics are R. A. Fisher, J. B. S. Haldane and S. Wright, and it is appropriate to consider briefly the general conclusions reached by these men regarding the most important evolutionary forces. Fisher, to whom many of the results in this book are due, argued strongly in favour of the view that, given the mutational building-blocks, selection is the only significant force acting in evolution. Thus Fisher, strongly influenced by results such as (6.1), would argue that, apart from the special case of the problem of survival of new mutants, the effect of random fluctuations in gene frequencies can be almost entirely ignored. This view is, of course, reinforced by the argument that normally it is the large population rather than the small population that is important in evolution.

Wright, who is again responsible for many of the results reached above, would attribute rather more importance to chance factors than would Fisher. Wright's view is that most species contain a number of small subpopulations which are sufficiently isolated to allow different frequencies for the various genes in different subpopulations, but not so isolated that favourable genes cannot spread eventually through all subpopulations. Within each subpopulation there will be a number of stationary points (typified by (9.18)). The set of gene frequencies will drift randomly in the neighbourhood of such a point until, either by random drift or through slight changes in the selective values themselves, they will move to a point where selective forces direct gene frequencies to some other equilibrium point. This process is then repeated indefinitely. Thus the role of random sampling is seen mainly as a trigger mechanism, initiating variation which is subsequently directed by natural selection. Under suitable conditions, this change in gene frequencies then spreads (through migration) into all subpopulations. Haldane's views on this matter would probably have been somewhere between these extremes.

It would be inappropriate to discuss mathematically the relative merits of these viewpoints, since the real questions are only to a minor

K

extent mathematical, although there are some mathematical points we have made which bear repeating. Thus the discussion of two-locus systems makes it clear that the gene frequencies in a population do not necessarily move under natural selection so as to maximize fitness. To this extent one should not attribute to natural selection a uniformly favourable character. Another point is that Section 4.5 suggests very strongly that population subdivision alone (unaccompanied by different sets of selective values in different subpopulations) is of very minor importance, and it might well be the case that populations which are sufficiently isolated to allow different gene frequencies in different subpopulations but not so isolated that favoured genes cannot spread with reasonable speed do not exist to any significant extent in nature. If different sets of selective values occur in different subpopulations, the term 'favoured gene' could lose all meaning and migration from one subpopulation to another might well have no significant effect on gene frequencies.

On the whole it appears to this author that, so far as a mathematical treatment is relevant, the view that selective forces are by far the most important factors in evolution is justified; (this does not, I believe, contradict the views of Wright). Certainly the effect of mutation, other than providing the genetical raw material, is completely negligible in determining gene frequencies. Altogether these general conclusions support strongly the evolutionary principles and theories laid down a century ago by Darwin.

# Bibliography

BODMER, W. F., 1964. 'Differential fertility in population genetics models.' *Genetics*, **51**, 411–4.

BODMER, W. F. and PARSONS, P. A., 1962. 'Linkage and recombination in evolution.' *Advan. Genet.*, **11**, 1–100.

CHIA, A. B., 1966. 'Some problems in mathematical genetics.' Unpublished M.Sc. Thesis. Australian National University.

CLARK, B., 1964. 'Frequency-dependent selection for the dominance of rare polymorphic genes.' *Evolution*, **18**, 364–9.

CROSBY, J. L., 1963. 'The evolution and nature of dominance.' *J. Theoret. Biol.*, **5**, 35–51.

CROW, J. F., 1958. 'Some possibilities for measuring selection intensities in man.' *Human Biology*, **30**, 1–13.

CROW, J. F. and KIMURA, M., 1956. 'Some genetic problems in natural populations.' *Proc. 3rd Berkeley Symp. Math. Stat. and Prob.* Vol. IV, 1–22, U. Calif. Press.

CROW, J. F. and KIMURA, M., 1963. 'The measurement of effective population number.' *Evolution*, **17**, 279–88.

CROW, J. F. and KIMURA, M., 1964. 'The number of alleles that can be maintained in a finite population.' *Genetics*, **49**, 725–38.

EWENS, W. J., 1964a. 'Correcting diffusion approximations in finite genetic models. Stanford University Technical Report, Department of Mathematics.

EWENS, W. J., 1964b. 'The pseudo-transient distribution and its uses in genetics.' *J. Appl. Prob.*, **1**, 141–56.

EWENS, W. J., 1965. 'The adequacy of the diffusion approximation to certain distributions in genetics.' *Biometrics*, **21**, 386–94.

EWENS, W. J., 1967. 'A note on the mathematical theory of the evolution of dominance.' *Amer. Nat.*, **101**, 35–40.

EWENS, W. J. 1968a.'A genetic model having complex linkage behaviour.' To appear in *Der Zuchter* by J. Theoret, and *Appl. Genetics*, **38**, 140–3.

EWENS, W. J., 1968b. 'Linkage, epistasis, and the genetic load.' Submitted to *Genet. Research*.

EWENS, W. J. and EWENS, P. M., 1966. 'The maintenance of alleles by mutation—Monte Carlo results for normal and self-sterility populations' *Heredity*, **21**, 371–8.

FELLER, W., 1954. 'Diffusion processes in one dimension.' *Trans. Amer. Math. Soc.*, **77**, 1–31.

FELSENSTEIN, J., 1965. 'The effect of linkage on directional selection.' *Genetics*, **52**, 349–63.

FISHER, R. A., 1922. 'On the dominance ratio.' *Proc. Roy. Soc. Edinb.*, **42**, 321–41.

FISHER, R. A., 1928a. 'The possible modification of the response of the wild type to recurrent mutation.' *Amer. Nat.*, **62**, 115–26.

FISHER, R. A., 1928b. 'Two further notes on the origin of dominance.' *Amer. Nat.*, **62**, 571–4.

FISHER, R. A., 1930. '*The Genetical Theory of Natural Selection*. Clarendon Press, Oxford.

FISHER, R. A., 1939. 'Stage of development as a factor influencing the variance in the number of offspring, frequency of mutants and related quantities.' *Ann. Eugen.*, **9**, 406–8.

FISHER, R. A., 1941. 'Average excess and average effect of a gene substitution.' *Ann. Eugen.*, **11**, 53–63.

FISHER, R. A., 1958. *The Genetical Theory of Natural Selection*. (Second revised edition.) Dover, New York.

FORD, E. B. and SHEPPARD, P. M., 1965. 'Natural selection and the evolution of dominance.' *Heredity*, **21**, 139–46.

FRASER, G. R. (1962). 'Our genetical "load". A review of some aspects of genetical variation.' *Ann. Hum. Genet.*, **25**, 387–415.

HALDANE, J. B. S., 1924. 'A mathematical theory of natural and artificial selection.' *Trans. Camb. Phil. Soc.*, **23**, 19–41.

HALDANE, J. B. S., 1926. 'A mathematical theory of natural and artificial selection III.' *Proc. Camb. Phil. Soc.*, **23**, 363–72.

HALDANE, J. B. S., 1927a. 'A mathematical theory of natural and artificial selection IV.' *Proc. Camb. Phil. Soc.*, **23**, 607–15.

HALDANE, J. B. S., 1927b. 'A mathematical theory of natural and artificial selection V.' *Proc. Camb. Phil. Soc.*, **23**, 838–44.

HALDANE, J. B. S., 1930a. 'A mathematical theory of natural and artificial selection VI.' *Proc. Camb. Phil. Soc.*, **26**, 220–30.

HALDANE, J. B. S., 1930b. 'A mathematical theory of natural and artificial selection.' *Proc. Camb. Phil. Soc.*, **27**, 137–42.

HALDANE, J. B. S., 1932a. *The Causes of Evolution*. Longmans, Green.

HALDANE, J. B. S., 1932b. 'A mathematical theory of natural and artificial selection. IX. Rapid selection.' *Proc. Camb. Phil. Soc.*, **28**, 244–8.

HALDANE, J. B. S., 1939. 'The equilibrium between mutation and random extinction.' *Ann. Eugen.*, **9**, 400–5.

HALDANE, J. B. S., 1957. 'The cost of natural selection.' *J. Genet.*, **55**, 511–24.

HALDANE, J. B. S., 1963. 'The selection of double heterozygotes.' *J. Genet.*, **58**, 125–8.

HALDANE, J. B. S. and JAYAKAR, S. D., 1963. 'Polymorphism due to selection of varying direction.' *J. Genet.* **58**, Part 2, 237–242.

HARDY, G. H., 1908. 'Mendelian proportions in a mixed population.' *Science*, **28**, 49–50.

HARRIS, T. E., 1963. *The Theory of Branching Processes*. Springer, Berlin.

JAIN, S. K. and ALLARD, R. W., 1965. 'The nature and stability of equilibria under optimizing selection.' *Proc. Nat. Acad. Sci.*, **54**, 1436–43.

JAIN, S. K. and ALLARD, R. W., 1966. 'The effects of linkage, epistasis, and inbreeding on population changes under selection.' *Genetics*, **53**, 633–59.

KARLIN, S. and MCGREGOR, J. L., 1962. 'On a genetic model of Moran.' *Proc. Comb. Phil. Soc.*, **58**, Part 2, 299–311.

KARLIN, S. and MCGREGOR, J. L., 1964. 'Direct product branching processes and related Markov chains.' *Proc. Nat. Acad. Sci.*, **51**, 598–602.

KARLIN, S. and MCGREGOR, J. L., 1965. 'Direct product branching processes and related induced Markov chains. 1. Calculation of rates of approach to homozygosity. In *Bernoulli (1723 Bayes (1963) Laplace (1813) Anniversary Volume*, ed. L. M. LeCam and J. Neyman. Springer.

KARLIN, S. and MCGREGOR, L. J., 1966. 'The number of mutant forms maintained in a population.' Stanford University Technical Report, Department of Mathematics.

KARLIN, S., MCGREGOR, J. L. and BODMER, W. F., 1966. 'The rate of production of recombinations between linked genes in finite populations.' Stanford University Technical Report, Departments of Mathematics and Genetics.

KHAZANIE, R. G. and MCKEAN, H. E., 1964. 'A Mendelian Markov process with multinomial transition probabilities. 1. The binomial case.' Purdue University Technical Report, Department of Statistics.

KIMURA, M., 1955a. 'Process of random genetic drift with a continuous model.' *Proc. Nat. Acad. Sic.*, **41**, 144–50.

KIMURA, M., 1955b. 'Random genetic drift in multi-allelic locus.' *Evolution*, **9**, 419–35.

KIMURA, M., 1955c. 'Stochastic processes and gene frequencies.' *Cold. Spr. Harb. Symp. Quant. Biol.* Vol XX. Population Genetics, 35–41.

KIMURA, M., 1956. 'A model of a genetic system which leads to closer linkage by natural selection.' *Evolution*, **3**, 278–87.

KIMURA, M., 1957. 'Some problems of stochastic processes in genetics.' *Ann. Math. Statist.*, **28**, 882–901.

KIMURA, M., 1958. 'On the change of population fitness by natural selection. *Heredity*, **12**, 145–67.

137

KIMURA, M., 1960. 'Optimum mutation rate and degree of dominance as determined by the principle of minimum genetic load.' *J. Genet.*, **57**, 21–34.

KIMURA, M., 1964. 'Diffusion models in population genetics.' *J. Appl. Prob.*, **1**, 177–232.

KIMURA, M., 1965. 'Attainment of quasi linkage equilibrium when gene frequencies are changing by natural selection. *Genetics*, **52**, 875–90.

KINGMAN, J. F. C., 1961a. 'A mathematical problem in population genetics.' *Proc. Camb. Phil. Soc.*, **57**, 574–82.

KINGMAN, J. F. C., 1961b. 'A matrix inequality.' *Quart. J. Math.*, **12**, 78–80.

KOJIMA, K. and KELLEHER, T. M., 1962. 'Survival of mutant genes.' *Amer. Nat.*, Vol. XCVI, 329–46.

KOJIMA, K. and SCHAFFER, H. E., 1967. 'Survival process of linked mutant genes.' *Evolution*, **21**, 518–31.

LEVINS, R., 1965. 'Theory of fitness in a heterogeneous environment. V. Optical genetic systems.' *Genetics*, **52**, 891–904.

LEWIS, D., 1948. 'Structure of the incompatibility gene. 1. Spontaneous mutation rate.' *Heredity*, **2**, 219–36.

LEWONTIN, R. C., 1964a. 'The interaction of selection and linkage. 1. General considerations – heterotic models.' *Genetics*, **49**, 49–67.

LEWONTIN, R. C., 1964b. 'The interaction of selection and linkage. II. Optimum models.' *Genetics*, **50**, 757–82.

LEWONTIN, R. C. and HUBBY, J. L., 1966. 'A molecular approach to the study of genic heterozygosity in natural populations. II. Amount of variation and degree of heterozygosity in natural populations of *Drosophila pseudoobscura*.' *Genetics*, **54**, 595–609.

LEWONTIN, R. C. and KOJIMA, K., 1960. 'The evolutionary dynamics of complex polymorphisms.' *Evolution*, **14**, 458–72.

LI, C. C., 1963. 'Equilibrium under differential selection in the sexes.' *Evolution*, **17**, 493–6.

MANDEL, S. P. H., 1958. 'The stability of a multiple allelic system.' *Heredity*, **13**, 289–302.

MATHER, K., 1941. 'Variation and selection of polygenic characters.' *J. Genet.*, **41**, 159–93.

MAYO, O., 1966. 'On the evolution of dominance.' *Heredity*, **21**, 499–511.

MORAN, P. A. P., 1958. 'Random processes in genetics.' *Proc. Camb. Phil. Soc.*, **54**, 60–71.

MORAN, P. A. P., 1960. 'The survival of a mutant gene under selection, II.' *J. Aust. Math. Soc.*, **1**, 485–91.

MORAN, P. A. P., 1962. *The Statistical Processes of Evolutionary Theory*. Clarendon Press, Oxford.

MORAN, P. A. P., 1964. 'On the nonexistence of adaptive topographies.' *Ann. Hum. Genet.*, **27**, 383–93.

MORAN, P. A. P., 1965. 'Unsolved problems in evolutionary theory.' *Proc. 5th Berk. Symp. Math. Stat. and Prob.* (To appear.)

MORTON, N. E., CROW, J. F. and MULLER, H. J., 1956. 'An estimate of

the mutational damage in man from data on consanguineous marriages.'
*Proc. Nat. Acad. Sci.*, **42**, 855–63.

NEI, M., 1965. 'Effect of linkage on the genetic load manifested under
inbreeding.' *Genetics*, **51**, 679–88.

OWEN, A. R. G., 1953. 'A genetical system admitting of two distinct stable
equilibria under natural selection.' *Heredity*, **7**, 97–102.

O'DONALD, P., 1967a. 'On the evolution of dominance, over-dominance
and balanced polymorphism.' *Proc. Roy. Soc. B.*, **268**, 216–28.

O'DONALD, P., 1967b. 'The evolution of selective advantage in a
deleterious mutation.' *Genetics*, **56**, 399–404.

PARSONS, P. A., 1963. 'Complex polymorphisms where the coupling and
repulsion double heterozygotes differ.' *Heredity*, **18**, 369–74.

PEARSON, K., 1904. 'On a generalized theory of alternative inheritance,
with special references to Mendel's laws.' *Phil. Trans. Roy. Soc. A*, **203**,
53–86.

POLLAK, E., 1966. 'On the survival of a gene in a sub-divided population.'
*J. Appl. Prob.*, **3**, 142–55.

ROBERTSON, A., 1962. 'Selection for heterozygotes in small populations.'
*Genetics*, **47**, 1291–1300.

SMITH, S. M., 1955. Appendix to 'Notes on sickle-cell polymorphism', by
A. C. Allison. *Ann. Hum. Genet.*, **19**, 51–57.

TURNER, J. R. G., 1967. 'On supergenes. I. The evolution of supergenes.'
*Amer. Nat.*, **101**, 195–221.

WATTERSON, G. A., 1962. 'Some theoretical aspects of diffusion theory in
population genetics.' *Annals of Math. Stat.*, **33**, 939–57.

WATTERSON, G. A., 1964. 'The application of diffusion theory to two
population genetic models of Moran.' *J. Appl. Prob.*, **1**, 233–46.

WEINBERG, W., 1908. 'Über den Nachweis der Vererbung beim
Menschen.' *Jh. Ver. Vaterl. Naturk. Wurttemb.*, **64**, 368–82.

WHITTAKER, E. T. and WATSON, G. N., 1962. *A Course in Modern
Analysis* (fourth edit.) Cambridge University Press.

WRIGHT, S., 1931. 'Evolution in Mendelian populations.' *Genetics*, **16**,
97–159.

WRIGHT, S., 1937. 'The distribution of gene frequencies in populations.'
*Proc. Nat. Acad. Sci. Wash.*, **23**, 307–20.

WRIGHT, S., 1942. 'Statistical genetics and evolution.' *Bull Amer. Math.
Soc.*, **48**, 223–46.

WRIGHT, S., 1945. 'The differential equation of the distribution of gene
frequencies.' *Proc. Nat. Acad. Sci. Wash.*, **31**, 382–9.

WRIGHT, S., 1948. 'On the roles of directed and random changes in gene
frequency in the genetics of populations.' *Evolution*, **2**, 279–94.

WRIGHT, S., 1949. 'Adaptation and selection. In: *Genetics, Paleontology
and Evolution*, Chap. 20, pp. 365–89. Edited by G. L. Jepson, G. G.
Simpson and E. Mayr. Princeton University Press.

WRIGHT, S., 1955. 'Classification of the factors of evolution.' *Cold Spr.
Harb. Symp. Quant. Biol.*, XX, 16–24.

WRIGHT, S., 1964. 'Pleiotrophy in the evolution of structural reduction
and dominance.' *Amer. Nat.*, XCVIII, 65–69.

# Glossary

Alleles. Two or more genes are alleles of each other if they can occupy the same locus on homologous chromosomes and can mutate to each other.

Chromosomes. Thread-shaped bodies, occurring in cell nuclei, along which genes are arranged in linear order.

Diploid. Having pairs of chromosomes in each nucleus, the members of each pair being homologous.

Dioecious. A population is dioecious if any given individual in it can act only as a male or a female.

Dominant. An allele which produces the same character when present with a certain other allele as when present at both loci is dominant to that allele.

Epistasis. Two loci are epistatic if the character expressed by a given gene at one locus depends on the gene present at the second locus.

Gamete. A reproductive haploid cell which fuses with another gamete to form a zygote.

Gene. A minute zone of a chromosome which is the fundamental unit of heredity. A gene partially or wholly governs the expression of a certain character or characters in an individual.

Genotype. The genetic constitution of an individual (as contrasted with the manifested characteristic or phenotype).

Haploid. Having a single set of chromosomes (one from each homologous pair) in each cell nucleus.

Heterozygote. Having two different alleles at corresponding loci of a pair of homologous chromosomes.

Homologous chromosomes. Chromosomes which contain identical sets of loci.

Homozygote. Having the same alleles at corresponding loci of a pair of homologous chromosomes.

Linkage. The tendency for genes which are physically close on the same chromosome to be passed on together.

Locus. A position on a chromosome at which exactly one gene must occur.

Monoecious. (of plants) having both male and female reproductive organs in the same individual; (of animals) producing both male and female gametes.

Population. An interbreeding group of individuals.

Recessive. The converse of dominant.

Recombination fraction. The recombination fraction between two given loci is the probability that two genes, one at each locus on the given chromosome, are not passed on together to an offspring.

Selection. Selection occurs when, due to differential fitnesses, certain genes have higher probabilities of transmission to future generations than others.

Sex chromosomes. Chromosomes of which there is a homologous pair in the cell nuclei of one sex (the homogametic sex) and a dissimilar pair, (one of which is similar to those in the homogametic sex) in the cell of nuclei of the other (heterogametic) sex.

Sex-linked genes. Genes on the sex chromosome.

Zygote. A diploid cell formed by the union of two gametes, and which develops into a new individual.

# Indexes

# Author Index

# Subject Index